A Reference Guide to Vector Algebra

A Reference Guide to Vector Algebra

T. P. Humphreys, PhD.
Science Tap, Inc. (formerly ToolChest Software, Inc.)

Jain Publishing Company, Inc.
Fremont, California

---

jainpub.com

Jain Publishing Company, Inc. is a diversified publisher of college textbooks and supplements, as well as professional and scholarly references, and books for the general reader. A complete, up-to-date listing of all the books, with cover images, descriptions, review excerpts, specifications and prices is always available on-line at **jainpub.com**. Our booksPLUS® division provides custom publishing and related services in print as well as electronic formats. Our learn24x7® division provides e-learning related products and services.

Printed in USA

ISBN 978-0-87573-095-0

Preface

For the Student:
This handbook is written as an introductory reference book in vector algebra and as a companion guide to solving vector algebra problems using the accompanying Vector Algebra Tools software. Each chapter begins by exploring a particular vector algebra topic and outlines problem solving applications in geometry, physics and engineering. In addition, numerous worked examples of varying difficulty are included to augment your understanding of each vector algebra topic and to develop your problem solving skills through the use of the Vector Algebra Tools. Therefore, whether you are involved in classroom or independent study assignments, preparing for a test, or solving a specific problem, *A Reference Guide to Vector Algebra* will help you acquire the necessary problem solving skills, strategies, and confidence to solve a wide variety of vector algebra problems.

For the Instructor:
The level of presentation and content of this handbook is intended to complement AP high-school and first year college curriculum topics in vector algebra and can be used as a quick and convenient reference guide to vector algebra concepts, methods, laws and problem solving applications. *A Reference Guide to Vector Algebra* can be used in a variety of teaching environments including classroom presentations, test question composition, devising problem solving strategies and one-on-one tutoring sessions.

Accompanying Software on CD-ROM:
The Vector Algebra Tools by Science Tap, Inc. (formerly ToolChest Software, Inc.) are a comprehensive set of specialized vector calculators that facilitates the algebraic determination of vector quantities in a Cartesian coordinate system. The software is specifically designed to operate on Microsoft® Windows® 7, Vista, XP and 2000 operating systems.

Second Edition
This second edition contains changes and improvements in response to comments by readers. In particular, some sections and problems have been rewritten to improve clarity and understanding. Additional worked examples have been included in the book and in the Vector Algebra Tools software. Lastly, the software has been modified to operate on Microsoft® Windows® 7.

Acknowledgements

I wish to thank my wife Elaine for her steadfast enthusiasm, support and encouragement, and without whose help and insight would not have made a creative idea become a reality. In addition, I extend a special thanks to Nancy Dawkins and Elizabeth Dawkins for many helpful suggestions and proof reading the manuscript. I would also like to thank Mukesh Jain from Jain Publishing Company, Inc., for his whole-hearted support of this second edition and who, when approached to publish this book with accompanying software, did not hesitate to do so.

Contents

CHAPTER 1

VECTOR ARITHMETIC

This chapter introduces the physical quantities of scalars and vectors. In particular, vector notation, definitions and the basic operations of vector arithmetic, including vector addition, subtraction and the scalar multiplication of a vector, are presented. Numerous problem solving applications and worked examples are included. Also demonstrated is the use of the Vector Arithmetic software tool in solving the example problems.

1.1 DEFINITIONS OF PHYSICAL QUANTITIES AND BASIC NOTATION

There are two kinds of physical quantities, scalars and vectors.

1.1.1 Scalar quantities

Scalar quantities, or **scalars**, are physical quantities that have a magnitude and a unit of measurement but no associated direction. Since scalar quantities are represented by real numbers, they obey the fundamental laws of elementary algebra and can be added, subtracted, multiplied and divided. There are numerous examples of scalar quantities in the physical world. These include mass, distance, time, speed, work, power, energy, heat, temperature, volume, and electric and magnetic flux. For example, the scalar quantities for mass and time are denoted by the letters m and t, respectively, with kilogram and second as their corresponding International System of Units (abbreviated SI). See Appendix E for a list of commonly used SI units.

1.1.2 Vector quantities

Vector quantities, or **vectors**, are physical quantities that have both a magnitude and a direction. Displacement, velocity, acceleration, force, linear momentum, angular momentum, torque, and electric and magnetic field intensities are all measurable vector quantities. For example, the displacement vector can be represented graphically as a **directed line segment**, where the vector $\overrightarrow{\mathrm{RS}}$, which joins the initial point R to the terminal point S, is represented by the vector **a**, as shown in figure 1.1. The length of the line drawn to scale indicates the vector's magnitude and the line's arrowhead indicates that the direction of **a** is along the line segment from R to S. Specifically, the magnitude of the displacement vector **a** is simply the distance between the two points R and S. Throughout this book, a vector, such as $\overline{\text{RS}}$, is represented in bold typeface, such as vector **a**, while its magnitude or modulus is denoted as | **a** | or simply as *a*. Consequently, knowing both the magnitude and direction of a physical quantity defines that quantity as a vector.

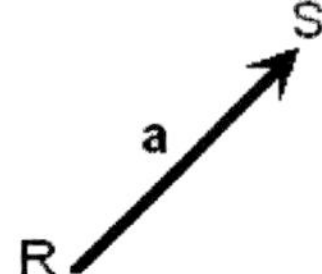

Fig. 1.1. The displacement of point S from point R can be represented by the vector **a**, where the distance between the two points is given by | **a** |.

In general, the definition of a vector makes no reference to its position; and therefore, vectors can be represented by a series of parallel straight lines of equal length and can be translated without changing their magnitude or direction. Vectors that are defined and illustrated in this manner have no specific line of action or point of application and are considered **free vectors**. By comparison, using a force vector as an example, the force acting on a rigid body has a specific line of action and, therefore, can be uniquely represented by a single **localized vector**. In this book no distinction will be made between free and localized vectors when performing vector algebra operations.

To represent a vector that is perpendicular to a two-dimensional plane, such as the page of this book, the diagram convention illustrated in figure 1.2 can be used. For example, if a vector points out from the page towards you, the vector can be represented by a circle with a dot at its center, where the dot represents the point or tip of the of vector's arrow moving towards you. In contrast, if a vector points into the page away from you, the vector can be represented by a circle with an enclosed cross, where the cross represents the vanes of the vector's arrow moving away from you. This method of representing a vector is generally used in the conjunction with the **right-hand rule** to indicate the direction of a perpendicular vector that results from the vector product of two vectors that are in the same plane, which, in our example, is the page of this book. Examples of various vector product quantities that can be represented in this manner include torque, angular velocity, angular momentum, and the forces on a moving electric charge or on a current carrying wire or loop, in a uniform external magnetic field.

Fig. 1.2. Magnetic field vectors **B** pointing (a) out of and (b) into the plane of the page.

1.2 THE RIGHT-HAND RULE

1.2.1 Construction of a right-handed three-dimensional Cartesian coordinate system

A **right-handed three-dimensional Cartesian coordinate system** is defined by a set of the mutually perpendicular x, y and z-axes, where the directions of the positive x, y and z-axes are governed by the right-hand rule. By convention, a right-handed three-dimensional Cartesian coordinate system is used routinely in solving elementary vector algebra problems in physics and engineering. Throughout this book the right-handed three-dimensional Cartesian coordinate system will be referred to simply as the three-dimensional Cartesian coordinate system.

To construct a three-dimensional Cartesian coordinate system, simply draw three mutually perpendicular lines Ox, Oy and Oz through the origin O to represent each axes. First establish that these axes are right-handed and orthogonal, i.e., perpendicular to each other, by applying the right-hand rule to each pair of axes. To apply the right-hand rule, hold your right hand along Oy and then curl your fingers inward from Oy to Oz. Your extended thumb should point along Ox. Likewise, hold your right hand along Oz and then curl your fingers inward from Oz to Ox. Your extended thumb should point along Oy. Now, hold your right hand along Ox and curl your fingers inward from Ox to Oy, so that your extended thumb points along Oz. Next draw the perpendiculars lines PT, PR, PS to the planes yOz, zOx, xOy, respectively, and complete the rectangular parallelepiped OASBCRPT, as illustrated in figure 1.3.

Referring to figure 1.3, the directions of the positive x, y and z-axes are represented by the **i**, **j** and **k** **unit vectors** (also referred to as the **Cartesian unit vectors**), respectively. These linearly independent unit vectors **i**, **j** and **k** are known as **base vectors** and form an orthonormal basis by which all vectors in a three-dimensional Cartesian coordinate system can be expressed. Hence, any vector **a** can be uniquely represented as a linear combination of the **i**, **j** and **k** unit vectors, and can be written in component form as

$$\mathbf{a} = a_x\,\mathbf{i} + a_y\,\mathbf{j} + a_z\,\mathbf{k},$$

where a_x, a_y and a_z are positive or negative numbers that correspond to the the x, y and z components of vector **a**, respectively. Vectors expressed in their component form are very useful in vector algebra enabling both a vector's magnitude and direction to be expressed simply in terms of its components. In addition, vector algebra operations, including the addition, subtraction and multiplication of vectors can be easily performed by algebraically manipulating the components of each vector.

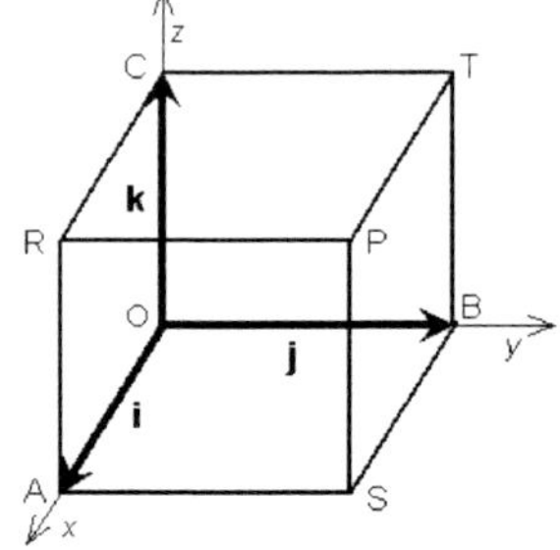

Fig. 1.3. Right-handed three-dimensional Cartesian coordinate system with the unit vectors **i**, **j** and **k** directed parallel to the positive x, y and z-axes, respectively.

1.2.2 Determination of the direction of a vector using the right-hand rule

The right-hand rule can also be used to determine the direction of a vector that results from the vector product of two vectors. To apply the right-hand rule to the vector product of any two nonzero vectors **a** and **b**, first extend your right hand and point your fingers to align with the direction of the x-axis or vector **a** referred to in figure 1.4. Then, keeping your thumb stationary, curl your fingers inwards towards the y-axis or vector **b**. Your thumb then points in the direction of the z-axis or the vector product vector **c**. This method can also be viewed as the rotation of a right-handed screw that advances in the direction of vector **c** when it is rotated in a counterclockwise direction from vector **a** to vector **b**. The vector product of two vectors and its method of calculation will be discussed in Chapter 5.

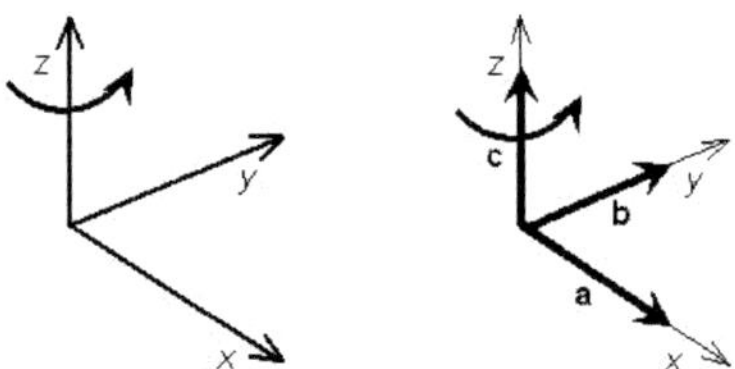

Fig. 1.4. Three-dimensional Cartesian coordinate system in which the coordinate x, y and z-axes and the vectors **a**, **b** and **c** satisfy the right-hand rule. Note that vector **c** is normal to the plane containing vectors **a** and **b**.

The right-hand rule is also used in the determination of the rotational direction of a magnetic field vector, **B**, of a current carrying wire, see figure 1.5, and is known as the **magnetic field right-hand rule**. To apply the magnetic field right-hand rule, you would hold the wire in your right hand so that your thumb points in the same direction as the conventional flow of current, I, i.e., the positive charge, and your fingers curl in the rotational direction of the magnetic field vector, **B**.

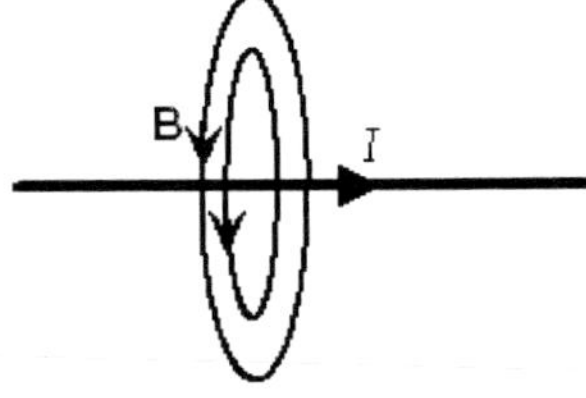

Fig. 1.5. The magnetic field **B** forms concentric circles around the current-carrying conductor, whose direction is determined by the magnetic field right-hand rule. The direction of the current, I, is determined by the flow of positive charge.

1.3 FURTHER VECTOR DEFINITIONS AND NOTATIONS

1.3.1 Negative vector

A **negative** or **inverse vector**, written, for example, as $-\mathbf{a}$, is a vector that has the same magnitude as vector **a** but points in the opposite direction, i.e., is **antiparallel** to vector **a**, as shown in figure 1.6.

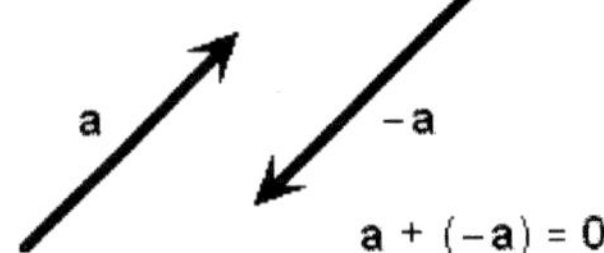

Fig. 1.6. The negative vector $-\mathbf{a}$ is antiparallel to vector **a**.

1.3.2 Zero or null vector

A **zero** or **null vector**, which is written as **0**, is a vector having a magnitude of zero and an indeterminate direction. For example, if vectors **a** and **b** are equal, then their vector difference results in a zero or null vector, such that $\mathbf{a} - \mathbf{b} = \mathbf{a} + (-\mathbf{b}) = \mathbf{0}$. Also, if the two points R and S in the directed line segment $\overrightarrow{\mathrm{RS}}$ coincide, then $\overrightarrow{\mathrm{RS}} = \mathbf{0}$. Furthermore, in a three-dimensional Cartesian coordinate system, a zero or null vector in component form is written as $\mathbf{0} = 0\,\mathbf{i} + 0\,\mathbf{j} + 0\,\mathbf{k}$, where **i**, **j** and **k** are the unit vectors in the direction of the positive x, y and z-axes, respectively.

1.3.3 Equal vectors

Vectors are said to be **equal** if they have the same magnitude and the same direction. For example, if $\mathbf{a} = \mathbf{b}$, then both vectors have equal magnitudes, i.e., $|\,\mathbf{a}\,| = |\,\mathbf{b}\,|$, and identical unit vectors indicating that they have the same direction. In addition, if **a** = **b**, then **a** − **b** = **0**. If the vectors **a** and **b** are expressed in component form, where **a** = a_x **i** + a_y **j** + a_z **k** and **b** = b_x **i** + b_y **j** + b_z **k**, then **a** = **b** if the components of both vectors are equal, such that $a_x = b_x$, $a_y = b_y$ and $a_z = b_z$. Note that if a vector is moved parallel to itself, the components of the vector are unaltered and the vector is unchanged since it has the same magnitude and direction.

1.3.4 Collinear vectors

Vectors are **collinear** if they lie on the same straight line or if they lie on parallel straight lines that can be moved onto the same straight line without changing their magnitudes and directions. See figure 1.7. For example, vectors **a** and **b** are collinear vectors and are linearly dependent, if one vector can be expressed as the scalar multiple of the other, such that **a** = λ**b**, where λ is a nonzero real number. When λ is positive, both **a** and **b** point in the same direction and are considered **parallel**. However, when λ is negative, both **a** and **b** point in opposite directions and are considered **antiparallel**. For λ = 1, both **a** and **b** are equal.

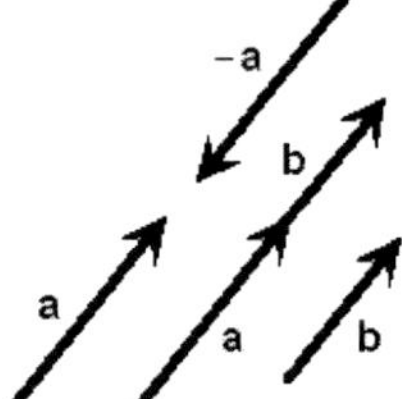

Fig. 1.7. Collinear vectors **a** and **b** that lie on the same straight line or parallel straight lines.

1.3.5 Coplanar vectors

Vectors that lie in a common plane or in parallel planes are known as **coplanar** vectors. Conversely, vectors that do not share the same plane or parallel planes are known as noncoplanar vectors. For example, the three linearly dependent nonzero vectors **a**, **b** and **c** are coplanar if each vector can be expressed as the linear combination of the other two vectors, such that

$$\lambda\mathbf{a} + \mu\mathbf{b} + \nu\mathbf{c} = \mathbf{0},$$

where λ, μ and ν are scalar multipliers that cannot all be zero. In particular, the three-dimensional vectors **a**, **b** and **c** are coplanar, if it can shown that that the scalar triple product of all three vectors is zero, i.e., **a** • (**b** × **c**) = 0. The determination of the scalar triple product of three vectors is discussed in Chapter 6.

1.3.6 Concurrent vectors

Vectors that have lines of action that meet or intersect at a common point are concurrent vectors. Conversely, vectors whose lines of action do not pass through the same point are called nonconcurrent vectors. Shown in figure 1.8 are two concurrent, coplanar force vectors $\mathbf{F}_1$ and $\mathbf{F}_2$, whose lines of action meet at the common point O in the same plane.

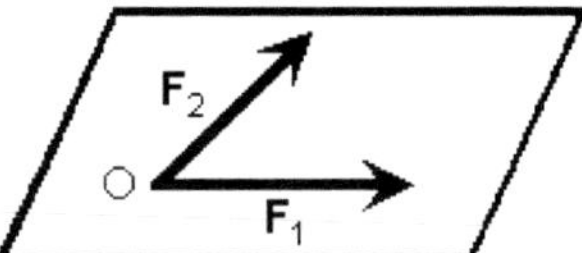

Fig. 1.8. Concurrent, coplanar force vectors $\mathbf{F}_1$ and $\mathbf{F}_2$ whose lines of action meet at a common point O.

1.3.7 Linear combinations of vectors in two and three dimensions

In two dimensions, vectors $\overrightarrow{OA}$ and $\overrightarrow{OB}$ are two noncollinear vectors that form the two sides of a parallelogram that are parallel to the vectors **a** and **b**, respectively, as shown in figure 1.9. Hence, vector **r** of any general point P, which is equal to $\overrightarrow{OP}$, can be expressed as the linear combination of the vectors $\overrightarrow{OA}$ and $\overrightarrow{OB}$, such that vector **r** is **coplanar** with the vectors **a** and **b** and is given by

$$\overrightarrow{OP} = \overrightarrow{OA} + \overrightarrow{OB};$$

$$\mathbf{r} = \lambda\mathbf{a} + \mu\mathbf{b},$$

where λ and μ are scalar multipliers. In this expression vectors **a** and **b** are **base vectors** and constitute a set of linearly independent vectors that can be used to express vector **r**. Accordingly, vectors $\lambda\mathbf{a}$ and $\mu\mathbf{b}$ are the **vector components** of vector **r** that are parallel to the base vectors **a** and **b**, respectively. The corresponding scalar multipliers λ and μ are the **components** of vector **r** along **a** and **b** as axes.

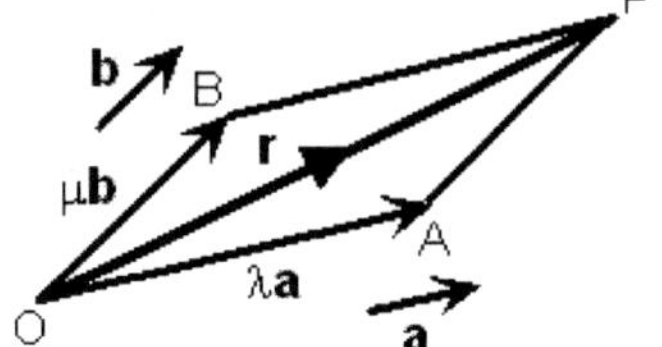

Fig. 1.9. Vector **r** can be expressed in two dimensions as $\mathbf{r} = \lambda\mathbf{a} + \mu\mathbf{b}$, where the given base vectors **a** and **b** are coplanar with **r**.

In three dimensions, any vector **r** can be expressed as the linear combination of three vectors that are parallel to the three noncoplanar base vectors, **a**, **b** and **c**. For example, if the vectors $\lambda\mathbf{a}$, $\mu\mathbf{b}$ and $\nu\mathbf{c}$ form the sides of a parallelepiped, as shown in figure 1.10, then the vector **r** of any general point P can be expressed as

$$\mathbf{r} = \lambda\mathbf{a} + \mu\mathbf{b} + \nu\mathbf{c},$$

where the vectors $\lambda\mathbf{a}$, $\mu\mathbf{b}$ and $\nu\mathbf{c}$ are the **vector components** of vector **r**, and the scalar multipliers λ, μ and ν are the **components** of vector **r** along **a**, **b** and **c** as axes, respectively. Note that the base vectors are linearly independent if it can be shown that $\lambda\mathbf{a} + \mu\mathbf{b} + \nu\mathbf{c} = \mathbf{0}$, which implies that $\lambda = \mu = \nu = 0$. However, this condition is met only if the base vectors **a**, **b** and **c** are noncoplanar.

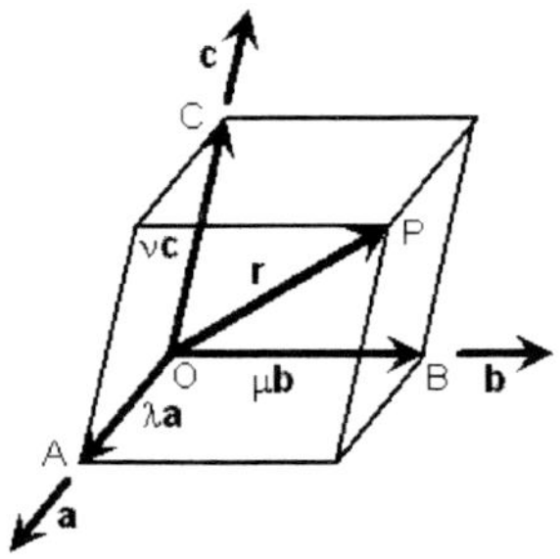

Fig. 1.10. Any vector **r** can be expressed in three-dimensions as $\mathbf{r} = \lambda\mathbf{a} + \mu\mathbf{b} + \nu\mathbf{c}$, where **a**, **b** and **c** are noncoplanar base vectors.

However, these general noncoplanar base vectors do not form an orthonormal basis, since they are neither mutually perpendicular to each other nor of unit length. Hence, the algebraic manipulations of vectors expressed in terms of these base vectors are rather cumbersome and lengthy. In elementary vector algebra, we choose instead the orthonormal base vectors, represented by the **i**, **j** and **k** unit vectors in a three-dimensional Cartesian coordinate system, to simplify calculations.

1.3.8 Position vector in a three-dimensional Cartesian coordinate system

In a three-dimensional Cartesian coordinate system, as shown in figure 1.11, a **position vector** or **radius vector a** that extends from the initial point O, defined at the origin, to the point P, whose location is given by the Cartesian coordinates (a_x, a_y, a_z), can be written in **component form** as

$$\mathbf{a} = a_x\,\mathbf{i} + a_y\,\mathbf{j} + a_z\,\mathbf{k}.$$

Specifically, position vector **a** is represented as a linear combination of the **i**, **j** and **k** unit vectors corresponding to the successive displacements of the vectors a_x **i**, a_y **j** and a_z **k** parallel to the positive x, y and z-axes, respectively. The vectors a_x **i**, a_y **j** and a_z **k** are the **Cartesian vector components** of vector **a**, more commonly referred to as the **vector components** of **a**. Therefore, vector **a** is said to be resolved into its **vector components** or **vector resolutes** resulting from the vector projections of vector **a** along the coordinate axes in the direction of the **i**, **j** and **k** unit vectors, such that

$$\mathbf{a} = (\mathbf{a} \bullet \mathbf{i})\,\mathbf{i} + (\mathbf{a} \bullet \mathbf{j})\,\mathbf{j} + (\mathbf{a} \bullet \mathbf{k})\,\mathbf{k}.$$

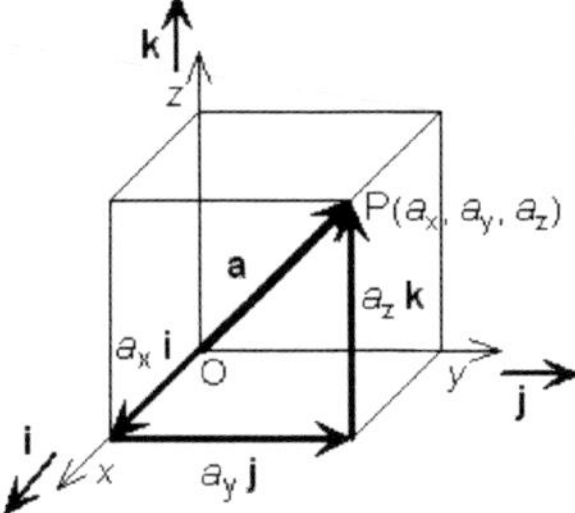

Fig. 1.11. Position vector **a** resolved into its vector components a_x **i**, a_y **j** and a_z **k** that are in the directions of the coordinate axes.

The **Cartesian components** of vector **a**, more commonly referred to as the **components** or **resolutes** of vector **a**, are the positive or negative numbers a_x, a_y and a_z that result from the scalar projections of vector **a** along the coordinate axes in the directions of the **i**, **j** and **k** unit vectors, such that

$$\mathbf{a} \bullet \mathbf{i} = (a_x\mathbf{i} + a_y\mathbf{j} + a_z\mathbf{k}) \bullet \mathbf{i} = a_x\mathbf{i} \bullet \mathbf{i} + a_y\mathbf{j} \bullet \mathbf{i} + a_z\mathbf{k} \bullet \mathbf{i} = a_x,$$

given that $\mathbf{i} \bullet \mathbf{i} = 1$ and $\mathbf{j} \bullet \mathbf{i} = \mathbf{k} \bullet \mathbf{i} = 0$. Similarly, we can write

$$\mathbf{a} \bullet \mathbf{j} = a_y \text{ and } \mathbf{a} \bullet \mathbf{k}. = a_z.$$

In particular, a_x is the x component of vector **a** along the positive x-axis in the direction of **i**, and a_y is the y component of vector **a** along the positive y-axis in the direction of **j**. Likewise, a_z is the z component of vector **a** along the positive z-axis in the direction of **k**. Thus, depending on the orientation that vector **a** makes with x, y and z coordinate axes, the components a_x, a_y and a_z may be positive or negative. For example, if a_x is positive, then the corresponding vector component a_x **i** points in the positive direction of the x-axis. Conversely, if a_x is negative, then the corresponding vector component $-a_x$ **i** points in the negative direction of the x-axis. Moreover, since the magnitude of a vector is always positive in value, we take the absolute values of the components $|\, a_x \,|$, $|\, a_y \,|$ and $|\, a_z \,|$ to represent the magnitudes of the corresponding vector components given by a_x **i**, a_y **j** and a_z **k**, respectively.

1.3.9 Magnitude of a vector

The **magnitude** or **modulus** of vector **a** is a scalar quantity, which is represented by a positive real number and, using vector **a** as an example, is written as | **a** | or simply as a. The magnitude of vector **a** corresponds to the length of vector **a** and is calculated algebraically from the components of **a** by using simple trigonometry based upon the Pythagorean theorem. Referring to figure 1.12 as an example, in a three-dimensional coordinate system, the magnitude of vector **a** can be determined geometrically by first calculating the magnitude of the vector $\overline{\mathrm{OS}}$, which forms the hypotenuse of the right angle triangle ORS in the x-y plane and is given by

$$|\,\overline{\mathrm{OS}}\,| = \sqrt{a_x^{\,2} + a_y^{\,2}}\,.$$

Then, the magnitude of vector **a** can be determined from the right angle triangle OSP, where $\mathbf{a} = \overrightarrow{\mathrm{OP}} = \overrightarrow{\mathrm{OS}} + \overrightarrow{\mathrm{SP}}$, and is given by

$$|\,\mathbf{a}\,| = a = |\,\overrightarrow{\mathrm{OP}}\,| = \sqrt{a_x^{\,2} + a_y^{\,2} + a_z^{\,2}}\,.$$

Therefore, we take the positive square root, since by definition the magnitude of a vector is always positive.

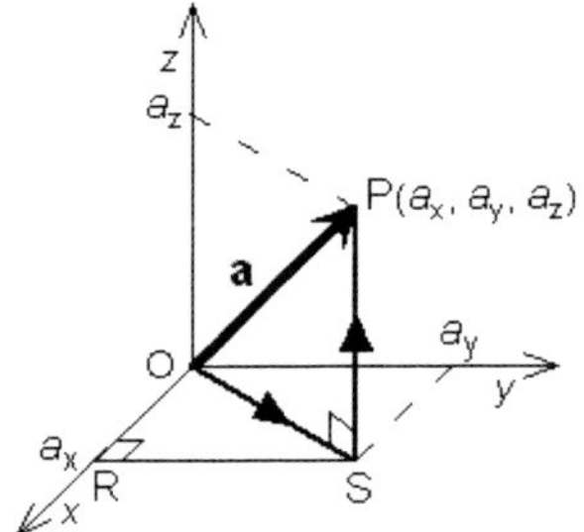

Fig. 1.12. Magnitude of vector **a** in a three-dimensional Cartesian coordinate is given by $|\,\mathbf{a}\,| = |\,\overrightarrow{\mathrm{OP}}\,| = \sqrt{a_x^{\,2} + a_y^{\,2} + a_z^{\,2}}\,.$

In a two-dimensional Cartesian coordinate system, if vector **a** is defined in the Cartesian plane xOy and corresponds to the vector $\overline{\mathrm{OS}}$, then vector **a** in component form can be written as

$$\mathbf{a} = a_x\,\mathbf{i} + a_y\,\mathbf{j},$$

and its magnitude can be expressed as

$$|\,\mathbf{a}\,| = a = \sqrt{a_x^{\,2} + a_y^{\,2}}\,.$$

1.3.10 Unit vector

A **unit vector**, or **normalized vector**, is a dimensionless vector that has a magnitude of one and a specific direction. In print unit vectors are distinguishable from other vectors as they are represented in bold typeface with a circumflex, e.g., $\hat{\mathbf{a}}$. The unit vector $\hat{\mathbf{a}}$, which specifies the direction of vector **a**, is given by the expression

$$\hat{\mathbf{a}} = \frac{\mathbf{a}}{|\,\mathbf{a}\,|},$$

where $|\,\mathbf{a}\,|$ is the magnitude of the nonzero vector **a**. Therefore, by dividing *any* vector by its magnitude yields a unit vector that is in the same direction as the vector. The magnitude of unit vector $\hat{\mathbf{a}}$ equals one. The process of creating a unit vector from a vector of arbitrary length is known as **normalizing** a vector. Therefore, it follows that any nonzero vector **a** can be determined from the product of its magnitude $|\,\mathbf{a}\,|$ and unit vector, $\hat{\mathbf{a}}$, and is given by the expression

$$\mathbf{a} = |\,\mathbf{a}\,|\,\hat{\mathbf{a}},$$

where unit vector $\hat{\mathbf{a}}$ has the same direction as vector **a**, as shown in figure 1.13. The synthesis of a vector from its unit vector is discussed in Chapter 2.

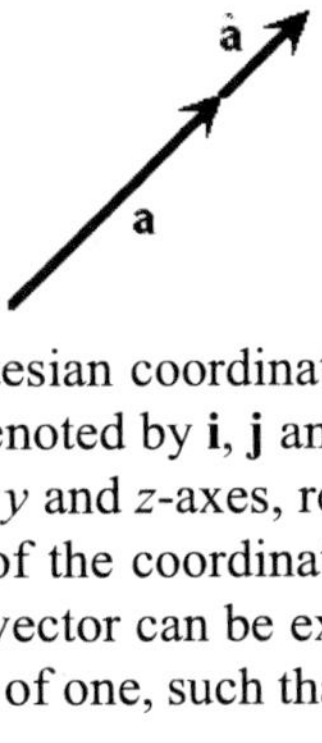

Fig. 1.13. The unit vector $\hat{\mathbf{a}}$ specifies the direction of vector **a**, such that $|\,\hat{\mathbf{a}}\,| = 1$.

As previously discussed, in a three-dimensional Cartesian coordinate system, the three mutually orthogonal unit vectors, denoted by **i**, **j** and **k**, are used to specify the directions of the positive *x*, *y* and *z*-axes, respectively. These unit vectors are the base vectors of the coordinate system and form an orthonormal basis in which any vector can be expressed. The **i**, **j** and **k** unit vectors have a magnitude of one, such that $|\,\mathbf{i}\,| = |\,\mathbf{j}\,| = |\,\mathbf{k}\,| = 1$.

1.4 VECTOR ARITHMETIC DETERMINATIONS

1.4.1 Graphical determination of the vector addition of two vectors

The single vector that results when two or more vectors of similar type, i.e., vectors that have the same units of measurements, are added together is called the **resultant vector** or **vector sum**. To determine the resultant vector of any two concurrent, coplanar vectors **a** and **b** graphically, we can use either the **triangle law** or the **parallelogram law** of vector addition, as shown in figure 1.14. To add vectors **a** and **b** using the triangle law, we place the tail of **b** at the head of **a** and draw the resultant vector **c** = **a** + **b** from the tail of **a** to the head of **b**, as shown in figure 1.14(a). Alternatively, using the parallelogram law, we place the tails of vectors **a** and **b** together and construct a parallelogram using vectors **a** and **b** as the sides of the parallelogram, as shown in figure 1.14(b). The vector that passes through the point where the tails of **a** and **b** intersect and is along the diagonal of the parallelogram represents the resultant vector **c** = **a** + **b**.

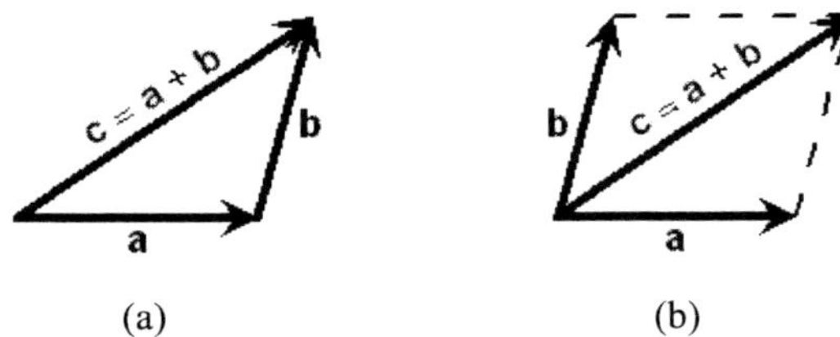

Fig. 1.14. Addition of vectors **a** and **b** using (a) the triangle law of vector addition and (b) the parallelogram law of vector addition to determine the resultant vector **c**, where **c** = **a** + **b**.

It is clear from figure 1.14 that both the triangle and parallelogram laws of vector addition are equivalent and either may be used to determine the resultant vector graphically. For example, the resultant force vector of two coplanar forces acting at a common point in the same plane can be determined using either the triangle or parallelogram laws of vector addition. However, to determine the vector addition of three or more vectors we employ the **polygon method** of vector addition. Further information on the graphical construction of the resultant vector that is based on the polygon method of vector addition in the Cartesian x-y plane is discussed in Chapter 7.

1.4.2 Graphical determination of the vector subtraction of two vectors

Vector subtraction is the **vector difference** between any two vectors **a** and **b** of similar type, and can be expressed as **a** − **b**. However, following the rules of scalar algebra, the vector difference may be also be written as **a** + (−**b**), as shown graphically in figure 1.15(a). This figure is constructed using the triangle law of vector addition by adding vector **a** to vector −**b**. The resultant vector is **c**, where **c** = **a** + (−**b**) = **a** − **b**. Additionally, the vector difference, **c**, which is vector **a** − **b**, when added to **b** results in the vector **a**, and is constructed by drawing vectors **a** and **b** from the same point, as shown in figure 1.15 (b).

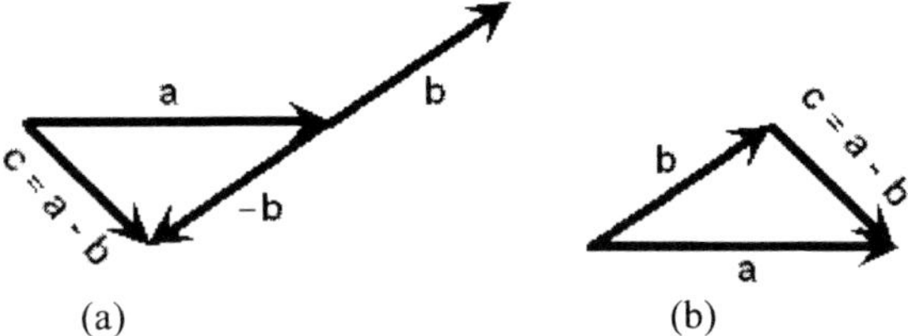

Fig. 1.15. Vector subtraction: (a) Using the triangle law of vector addition we add vector −**b** to vector **a**. (b) Alternatively, we can draw vectors **a** and **b** from the same point, where **c** = **a** − **b** is the vector that points from the head of **b** to the head of **a**.

These graphical methods of vector addition and subtraction are considered only approximations and do not yield accurate results. A more accurate method is to express each vector in its component form and then perform the operations algebraically, by manipulating the components of each vector. Moreover, since there is no graphical method for multiplying two or more vectors, these operations can only be performed algebraically. Specifically, two vectors can be multiplied by solving for either the scalar product or the vector product. Likewise, three vectors can be multiplied by solving for either the scalar triple product or vector triple product.

1.4.3 Algebraic determination of the addition and subtraction of two vectors in component form

If vectors **a** and **b** are represented in their component form in a three-dimensional Cartesian coordinate system, where $\mathbf{a} = a_x\,\mathbf{i} + a_y\,\mathbf{j} + a_z\,\mathbf{k}$ and $\mathbf{b} = b_x\,\mathbf{i} + b_y\,\mathbf{j} + b_z\,\mathbf{k}$, then the addition vector **r** and the subtraction vector **s** can be determined algebraically by adding or subtracting the components of the vectors as follows:

$$\mathbf{r} = \mathbf{a} + \mathbf{b} = (a_x\,\mathbf{i} + a_y\,\mathbf{j} + a_z\,\mathbf{k}) + (b_x\,\mathbf{i} + b_y\,\mathbf{j} + b_z\,\mathbf{k})$$

$$= (a_x + b_x)\,\mathbf{i} + (a_y + b_y)\,\mathbf{j} + (a_z + b_z)\,\mathbf{k},$$

where the components of **r** are given by $r_x = a_x + b_x$, $r_y = a_y + b_y$ and $r_z = a_z + b_z$.

$$\mathbf{s} = \mathbf{a} - \mathbf{b} = (a_x\,\mathbf{i} + a_y\,\mathbf{j} + a_z\,\mathbf{k}) - (b_x\,\mathbf{i} + b_y\,\mathbf{j} + b_z\,\mathbf{k})$$

$$= (a_x - b_x)\,\mathbf{i} + (a_y - b_y)\,\mathbf{j} + (a_z - b_z)\,\mathbf{k},$$

where the components of **s** are given by $s_x = a_x - b_x$, $s_y = a_y - b_y$ and $s_z = a_z - b_z$.

To determine the addition vector **r**, we separately add the x, y and z components of vectors **a** and **b** that are in the directions of the **i**, **j** and **k** unit vectors. Likewise, to determine the vector difference **s**, we separately subtract the x, y and z components of vector **b** from the x, y and z components of vector **a**, which are in the directions of the **i**, **j** and **k** unit vectors. In both cases, careful consideration is given to the signs of the components of each vector. Recall that vector **s** can also be considered the vector addition of vectors **a** and −**b** and can be expressed as **s** = **a** + (−**b**). Thus, both vectors **r** and **s** are **vector sums** or **resultant vectors**. Clearly, this **component method** of adding vectors can be easily extended to the determination of the resultant vector of an arbitrary number of vectors.

The magnitudes of vectors **r** and **s** can be obtained from their components as follows:

$$|\,\mathbf{r}\,| = r = \sqrt{(a_x + b_x)^2 + (a_y + b_y)^2 + (a_z + b_z)^2} \quad \text{and}$$

$$|\,\mathbf{s}\,| = s = \sqrt{(a_x - b_x)^2 + (a_y - b_y)^2 + (a_z - b_z)^2}.$$

In a three-dimensional Cartesian coordinate system, if the vectors **a** and **b** are the position vectors of the points A(a_x, a_y, a_z) and B(b_x, b_y, b_z), respectively, then the position vector of point A with respective to point B is represented by the displacement vector **s**, where **s** = **a** – **b**, and whose magnitude, i.e., the distance between the points A and B, is given by | **s** | = | **a** – **b** |.

1.4.4 Multiplication of a vector by a scalar

The multiplication of vector **a** by a nonzero real number scalar, λ, results in the vector λ**a**, whose direction is determined by the sign of λ. For example, if λ > 0, then λ**a** is a vector that is **parallel** to vector **a**, pointing in the same direction as vector **a**. Conversely, if λ < 0, then λ**a** is **antiparallel** to vector **a**, as it points in the opposite direction of vector **a**. The magnitude of the vector λ**a** is equal to the product of the magnitude of **a** and the scalar quantity λ and is expressed as $|\lambda\mathbf{a}|$ or $|\lambda||\mathbf{a}|$. Note that both vector **a** and the scaled vector λ**a** are related through the operation known as **scaling**, such that **a** and λ**a** are **collinear vectors**. Conversely, nonzero vectors that are not collinear cannot be related through scaling.

In a three-dimensional Cartesian coordinate system, if vector **a** is written in component form as $a_x\,\mathbf{i} + a_y\,\mathbf{j} + a_z\,\mathbf{k}$, then scaling **a** by a nonzero real number scalar λ gives the following vector:

$$\lambda\mathbf{a} = \lambda(a_x\,\mathbf{i} + a_y\,\mathbf{j} + a_z\,\mathbf{k}) = \lambda a_x\,\mathbf{i} + \lambda a_y\,\mathbf{j} + \lambda a_z\,\mathbf{k}.$$

1.4.5 Determination of the direction angles and the direction cosines of a vector

The direction of vector **r** in a three-dimensional Cartesian coordinate system is specified by the angles α, β and γ that vector **r** makes with the positive *x*, *y* and *z*-axes, respectively, as shown in figure 1.16. In this book the angles α, β and γ are referred to as the **direction angles** of vector **r**.

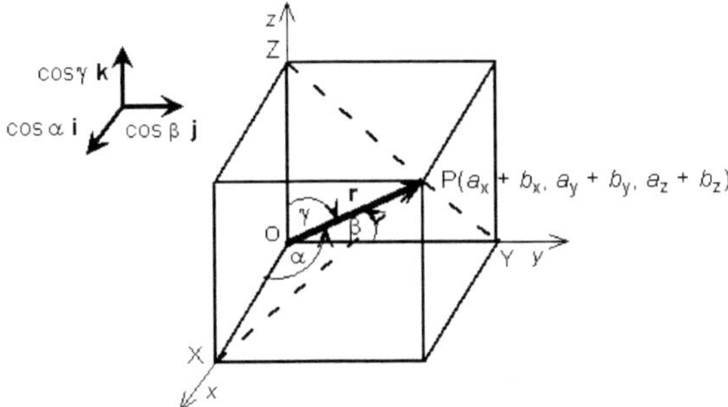

Fig. 1.16. The direction angles of vector **r** are the angles that vector **r** makes with each of the positive directions of the coordinate axes.

Given that $\mathbf{r} = (a_x + b_x)\,\mathbf{i} + (a_y + b_y)\,\mathbf{j} + (a_z + b_z)\,\mathbf{k}$, the direction angles between vector **r** and the positive coordinate axes can be determined from elementary trigonometry. Therefore, by referring to figure 1.16, triangle OXP is a right angle triangle with the right angle located at the point X at a distance $a_x + b_x$ from O. Hence, the angle α between

r and the positive x-axis is given by the expression $\cos\alpha = \dfrac{a_x + b_x}{|\mathbf{r}|}$, where $|\mathbf{r}|$ is the magnitude of vector **r**. Likewise, the triangles OYP and OZP are also right angle triangles. For the triangle OYP, the right angle is located at point Y at a distance $a_y + b_y$ from O; and, for the triangle OZP, the right angle is located at point Z at a distance $a_z + b_z$ from O. Similarly, the angles between vector **r** and the positive y and z-axes are given by $\cos\beta = \dfrac{a_y + b_y}{|\mathbf{r}|}$ and $\cos\gamma = \dfrac{a_z + b_z}{|\mathbf{r}|}$, respectively.

In summary, the direction angles α, β and γ of vector **r** can be determined using the following equations:

$$\cos\alpha = \frac{a_x + b_x}{|\mathbf{r}|},\ \cos\beta = \frac{a_y + b_y}{|\mathbf{r}|} \text{ and } \cos\gamma = \frac{a_z + b_z}{|\mathbf{r}|},$$

where the cosines of the direction angles are called the **direction cosines** of vector **r**. The direction cosines of **r** can also be denoted as $[\ell, m, n]$, where $\ell = \cos\alpha$, $m = \cos\beta$ and $n = \cos\gamma$. Specifically, the direction cosines of vector **r** are the components of the unit vector $\hat{\mathbf{r}}$ that is in the direction of **r**. Therefore, vector **r** can be expressed as the product of its magnitude, $|\mathbf{r}|$, and unit vector $\hat{\mathbf{r}}$, such that

$$\mathbf{r} = |\mathbf{r}|\,\hat{\mathbf{r}} = |\mathbf{r}|\,(\cos\alpha\,\mathbf{i} + \cos\beta\,\mathbf{j} + \cos\gamma\,\mathbf{k}) = |\mathbf{r}|\,(\ell\,\mathbf{i} + m\,\mathbf{j} + n\,\mathbf{k}),$$

where $|\mathbf{r}| = \sqrt{(a_x + b_x)^2 + (a_y + b_y)^2 + (a_z + b_z)^2}$ and $(\ell\,\mathbf{i} + m\,\mathbf{j} + n\,\mathbf{k})$ is the unit vector $\hat{\mathbf{r}}$. Since the components of a unit vector are the direction cosines of the vector, it is common practice to express a vector's direction in terms of its direction cosines rather than its direction angles.

If vectors **a** and **b** are defined in a two-dimensional Cartesian coordinate system, where $\mathbf{a} = a_x\,\mathbf{i} + a_y\,\mathbf{j}$ and $\mathbf{b} = b_x\,\mathbf{i} + b_y\,\mathbf{j}$, then **r** is defined in the xOy plane, given by $\mathbf{r} = (a_x + b_x)\,\mathbf{i} + (a_y + b_y)\,\mathbf{j}$, and we have the corresponding results:

$$\cos\alpha = \frac{a_x + b_x}{|\mathbf{r}|} \text{ and } \cos\beta = \frac{a_y + b_y}{|\mathbf{r}|};$$

and

$$\mathbf{r} = |\mathbf{r}|\,\hat{\mathbf{r}} = |\mathbf{r}|\,(\cos\alpha\,\mathbf{i} + \cos\beta\,\mathbf{j}) = |\mathbf{r}|\,(\ell\,\mathbf{i} + m\,\mathbf{j}),$$

where the magnitude of vector **r**, $|\mathbf{r}|$, is given by

$$|\mathbf{r}| = \sqrt{(a_x + a_y)^2 + (b_x + b_y)^2}.$$

1.4.6 Determination of the angle between two vectors

The angle θ between the directions of two vectors **a** and **b** is given by

$$\theta = \cos^{-1}(\ell\ell' + mm' + nn'),$$

where $[\ell, m, n]$ are the direction cosines of vector **a** and $[\ell', m', n']$ are the direction cosines of vector **b**. For **parallel** vectors, $\theta = 0°$ and $\ell\ell' + mm' + nn' = 1$. For **orthogonal**, i.e., perpendicular vectors, $\theta = 90°$ and $\ell\ell' + mm' + nn' = 0$. However, rather that calculating the direction cosines for each individual vector, which is rather tedious, it is simpler to calculate the angle θ using either the scalar product or vector product method. These methods of calculations are discussed in Chapters 4 and 5, respectively.

1.5. LAWS AND PROPERTIES OF VECTOR ADDITION ALGEBRA AND THE SCALAR MULTIPLICATION OF A VECTOR

Summarized below are the important laws and properties pertaining to vector addition and the scalar multiplication of a vector for the nonzero vectors **a**, **b** and **c** and the nonzero scalar quantities p and q.

$\mathbf{a} + \mathbf{b} = \mathbf{b} + \mathbf{a}$	Commutative law of vector addition
$\mathbf{a} + (\mathbf{b} + \mathbf{c}) = (\mathbf{a} + \mathbf{b}) + \mathbf{c}$	Associative law of vector addition
$p(\mathbf{a} + \mathbf{b}) = p\mathbf{a} + p\mathbf{b}$	Distributive law of vector addition
$p\mathbf{a} = \mathbf{a}p$	Commutative law of scalar multiplication
$p(q\mathbf{a}) = (pq)\mathbf{a}$	Associative law of scalar multiplication
$(p + q)\mathbf{a} = p\mathbf{a} + q\mathbf{a}$	Distributive law of scalar multiplication

These axioms apply only to nonzero vectors and when the scalar quantities are nonzero real numbers, thereby permitting the manipulation of vector quantities in the same manner as ordinary algebraic equations.

1.6 PROBLEM SOLVING APPLICATIONS OF VECTOR ARITHMETIC

The following examples demonstrate the problem solving applications of vector arithmetic.

- When a boat crosses a flowing river or a person walks in a moving train or airplane, the boat or person's motion relative to the ground is determined by the addition of their velocity vectors. For example, the relative velocity of a boat crossing a river with respect to a stationary observer on the embankment, $\mathbf{v}_{BE}$, can be expressed as the vector addition of the velocity of the boat with respect to the water due to the boat's propulsion, $\mathbf{v}_{BW}$, and the velocity of the water with respect to the observer on the embankment due the river's current, $\mathbf{v}_{WE}$, and can be expressed as

$$\mathbf{v}_{BE} = \mathbf{v}_{BW} + \mathbf{v}_{WE}.$$

The relative velocity $\mathbf{v}_{BE}$ can be determined algebraically by adding the velocity vectors in their component form or graphically by using the triangle law of vector addition, as shown in figure 1.17.

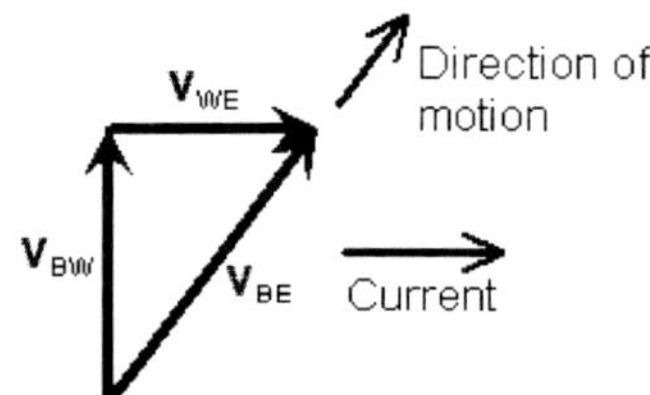

Fig. 1.17. The relative velocity of a boat from the point of view of an observer on the embankment is given by the vector $\mathbf{v}_{BE}$.

Note that a similar situation also arises when an aircraft flies into a cross wind such that the vector sum of the velocity of the aircraft with respect to the wind, $\mathbf{v}_{AW}$, and the velocity of the wind with respect to the ground, $\mathbf{v}_{WG}$, determines the relative velocity of the aircraft with respect to the ground, $\mathbf{v}_{AG}$, and is given by the expression

$$\mathbf{v}_{AG} = \mathbf{v}_{AW} + \mathbf{v}_{WG}.$$

- The relative velocity of an object moving with a constant velocity with respect to another object moving with a constant velocity is the vector difference of their velocities taken in the correct order. For example, if two race cars A and B are traveling with velocities $\mathbf{v}_{AG}$ and $\mathbf{v}_{BG}$ relative to the ground, respectively, then the relative velocity of car B with respect to car A, $\mathbf{v}_{BA}$, is given by the expression

$$\mathbf{v}_{BA} = \mathbf{v}_{BG} - \mathbf{v}_{AG}.$$

Thus, it follows that the relative velocity of car A with respect to car B is

$$\mathbf{v}_{AB} = \mathbf{v}_{AG} - \mathbf{v}_{BG}.$$

- Newton's second law of motion is expressed as $\mathbf{F}_{net} = \sum \mathbf{F} = m\mathbf{a}$, where $\mathbf{F}_{net}$ is the net external or resultant force acting on a particle of mass, m, that moves with a constant acceleration $\mathbf{a}$ in a straight line. The corresponding components for Newton's second law in a three-dimensional Cartesian coordinate system are expressed in terms of the x, y and z components of the resultant force as follows:

$$\sum F_x = ma_x, \ \sum F_y = ma_y \text{ and } \sum F_z = ma_z.$$

  Note that if $\mathbf{F}_{net} = 0$, then the particle is in **translational equilibrium** and the acceleration of the particle is zero.

- The linear momentum of a particle moving in a straight line is given by $\mathbf{p} = m\mathbf{v}$, where m is the mass of the particle and $\mathbf{v}$ is the particle's instantaneous velocity.

- The impulse of the average force $\mathbf{F}_{avg}$ acting on a particle during the time interval Δt is expressed as $\mathbf{I} = \mathbf{F}_{avg}\,\Delta t = \Delta\mathbf{p} = m(\mathbf{v}_2 - \mathbf{v}_1)$, where m is mass of the particle and $\mathbf{v}_1$ and $\mathbf{v}_2$ are the initial and final velocities of the particle, respectively. Note that the impulse $\mathbf{I}$ of the average force $\mathbf{F}_{avg}$ acting on the particle is equal to its change in momentum.

- In kinematics the equations of motion for a particle that moves in a straight line with a constant acceleration $\mathbf{a}$ are given by

$$\mathbf{v} = \mathbf{u} + \mathbf{a}t, \qquad \mathbf{s} = \mathbf{r} - \mathbf{r}_o = \mathbf{u}t + \tfrac{1}{2}\,\mathbf{a}t^2,$$

$$\mathbf{v}^2 = \mathbf{u}^2 + 2\mathbf{a}\mathbf{s}, \qquad \mathbf{s} = \left(\frac{\mathbf{u} + \mathbf{v}}{2}\right)t,$$

  where $\mathbf{u}$ and $\mathbf{v}$ are the initial and final instantaneous velocities, respectively, t is time and $\mathbf{r} - \mathbf{r}_o$ is the displacement vector $\mathbf{s}$ of the particle.

- The position vector $\mathbf{r}$ of the center of mass of a set of particles of masses, $m_1, m_2, m_3, \ldots, m_n$, situated respectively at the points with position vectors $\mathbf{r}_1, \mathbf{r}_2, \mathbf{r}_3, \ldots, \mathbf{r}_n$, is given by

$$\mathbf{r} = \frac{m_1\mathbf{r}_1 + m_2\mathbf{r}_2 + m_3\mathbf{r}_3 + \cdots}{m_1 + m_2 + m_3 + \cdots} = \frac{\sum m\mathbf{r}}{\sum m}.$$

The velocity of the center of mass of the set of particles, $\mathbf{v}_{cm}$, is given by

$$\mathbf{v}_{cm} = \frac{m_1\mathbf{v}_1 + m_2\mathbf{v}_2 + m_3\mathbf{v}_3 + \cdots}{m_1 + m_2 + m_3 + \cdots} = \frac{\sum m\mathbf{v}}{\sum m},$$

where $\mathbf{v}_1$, $\mathbf{v}_2$, $\mathbf{v}_3$, . . . , $\mathbf{v}_n$ are the velocity vectors of the masses m_1, m_2, m_3, . . . , m_n, respectively. Therefore, the total linear momentum, $\mathbf{p}_T$, of the system of particles is given by the expression

$$\mathbf{p}_T = \sum m\mathbf{v}_{cm} = M\mathbf{v}_{cm},$$

where M is the total mass of the system. The acceleration, of the center of mass of the set of particles, $\mathbf{a}_{cm}$, is given by

$$\mathbf{a}_{cm} = \frac{m_1\mathbf{a}_1 + m_2\mathbf{a}_2 + m_3\mathbf{a}_3 + \cdots}{m_1 + m_2 + m_3 + \cdots} = \frac{\sum m\mathbf{a}}{\sum m},$$

where $\mathbf{a}_1$, $\mathbf{a}_2$, $\mathbf{a}_3$, . . . , $\mathbf{a}_n$ are the acceleration vectors of the masses m_1, m_2, m_3, . . . , m_n, respectively. Therefore, from Newton's second law of motion, the net external force $\mathbf{F}_{net}$ acting on the system of particles is given by the expression

$$\mathbf{F}_{net} = \sum m\mathbf{a}_{cm} = M\mathbf{a}_{cm},$$

where M is the total mass of the system. If there are no resultant components of the net external force in the direction of each coordinate axes, then the **linear momentum** of the system is **conserved**. For example, given that

$$\sum F_x = \sum ma_x = 0,$$

the linear momentum of the system along the x-axis is constant. Similarly, if

$$\sum F_y = \sum ma_y = 0,$$

the linear momentum of the system along the y-axis is constant. Likewise, if

$$\sum F_z = \sum ma_z = 0,$$

the linear momentum of the system along the z-axis is constant.

- The external torque $\boldsymbol{\tau}$ acting on any rigid body that is rotating about a fixed axis through its center of mass is given by

$$\boldsymbol{\tau} = I\boldsymbol{\alpha},$$

where I is the moment of inertia about the axis of rotation and $\boldsymbol{\alpha}$ is the angular acceleration about the axis. Note that both $\boldsymbol{\tau}$ and $\boldsymbol{\alpha}$ point in the same direction along the axis of rotation.

- The total angular momentum of any rigid body that is rotating about a fixed axis is given by

$$\mathbf{L} = I\boldsymbol{\omega},$$

where I is the moment of inertia about the axis of rotation and $\boldsymbol{\omega}$ is the vector angular velocity about the axis. The direction of $\boldsymbol{\omega}$ is determined using the right-hand rule, such that, when the fingers of your right hand are curled in the direction of rotation, your extended thumb points in the direction of $\boldsymbol{\omega}$ that is along the axis of rotation. By convention a counterclockwise rotation corresponds to a positive angular velocity, and a clockwise rotation corresponds to a negative angular velocity. Note that both $\mathbf{L}$ and $\boldsymbol{\omega}$ point in the same direction along the axis of rotation.

- In a planar current loop of any shape with a vector area $\mathbf{A}$ and carrying a current I, the magnetic dipole moment $\boldsymbol{\mu}$ of the loop is given by

$$\boldsymbol{\mu} = I\mathbf{A},$$

where $\mathbf{A}$ is a vector normal to the plane of the loop and has a magnitude equal to the area of the loop. The direction of $\mathbf{A}$ is governed by the right-hand rule and is determined by curling the fingers of your right hand in the direction of the current in the loop, such that your upright thumb points in the direction of $\mathbf{A}$. Note that both $\boldsymbol{\mu}$ and $\mathbf{A}$ point in the same direction.

- The electric dipole moment $\mathbf{p}$ of an electric dipole consists of a pair of opposite charges of equal magnitude that are at a distance L apart, as shown in figure 1.18, and is given by

$$\mathbf{p} = |\,q\,|\,\mathbf{L},$$

where $|\, q \,|$ is the absolute magnitude of each charge and **L** is the displacement vector pointing from $-q$ to $+q$.

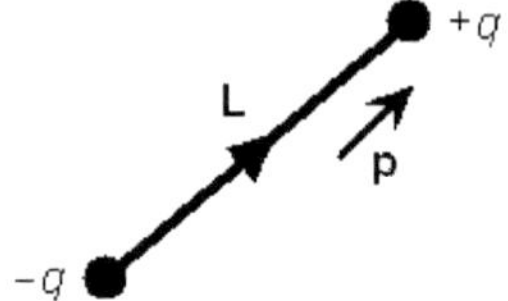

Fig. 1.18. Electric dipole moment **p** between a pair of opposite electric charges that are separated by a distance L, where $L = |\, \mathbf{L} \,|$.

- The force **F** on an electric charge q in a uniform external electric field **E** is given by $\mathbf{F} = q\mathbf{E}$, where the direction of **F** is determined by the sign of the charge q, as shown in figure 1.19. For example, a positively charged particle experiences a force that is in the same direction as **E**, and a negatively charged particle experiences a force that is in the opposite direction of **E**.

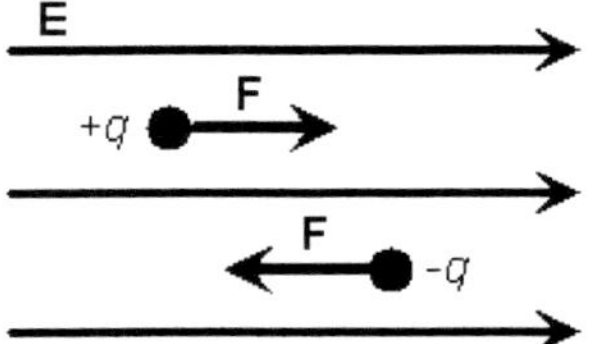

Fig. 1.19. Forces on opposite electric charges in a uniform electric field **E**.

- Ohm's law for a conductor is given by the expression $\mathbf{J} = \sigma\mathbf{E}$, where **J** is the electric current density at a point within the conductor, **E** is the electric field and σ is the conductivity of the conductor.

- In the classes of paramagnetic and diamagnetic materials, the magnetic state is defined by the magnetization vector **M** and is expressed as $\mathbf{M} = \chi\mathbf{H}$, where χ is a dimensionless quantity called the magnetic susceptibility of the medium and **H** is the magnetic field strength. For paramagnetic materials, where χ is positive, **M** is in the same direction as **H**. In contrast, for diamagnetic materials, where χ is negative, **M** is in the opposite direction of **H**.

- The vector equation of a straight line that passes through a given point A and is parallel to the vector **b** can be expressed as $\mathbf{r} = \mathbf{a} + \lambda\mathbf{b}$, where **r** is the position vector of any general point P on the straight line with respect to the origin at point O. See figure 1.20. The position vector of point A, with respect to the origin at point O, is given by **a**, and λ is a nonzero scalar multiplier. Hence, for each value of λ, the vector equation of the straight line gives the position vector of a point on the line.

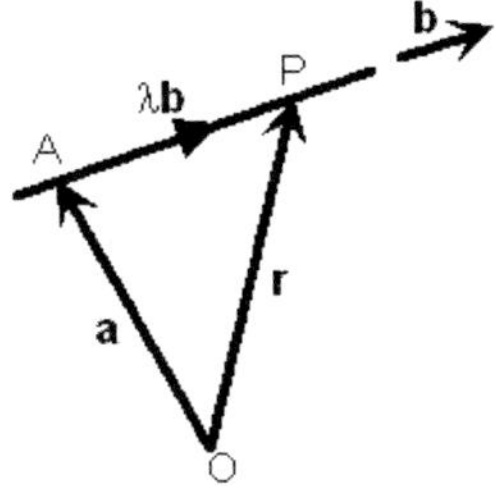

Fig. 1.20. The vector equation $\mathbf{r} = \mathbf{a} + \lambda\mathbf{b}$ of a straight line passing through the point A and parallel to vector **b**.

To determine for example, the two-dimensional vector equation of the straight line given by the Cartesian equation $2x + 3y = 18$, first calculate the position vector **a** for any general point that lies on the line. For instance, you can choose from the points A(3, 4), B(6, 2) and C (9, 0), all of which lie on the straight line. Let's choose the point A(3, 4), whose position vector from the origin is given by $\mathbf{a} = 3\,\mathbf{i} + 4\,\mathbf{j}$. Since the direction vector **b** is parallel to and in the direction of the straight line whose gradient is −2/3, the direction vector **b** is equivalent to the gradient of the line and can be written as $\mathbf{b} = 3\,\mathbf{i} - 2\,\mathbf{j}$. Therefore, the vector equation of the straight line can be expressed as

$$\begin{aligned} \mathbf{r} &= \mathbf{a} + \lambda\mathbf{b} \\ &= 3\,\mathbf{i} + 4\,\mathbf{j} + \lambda(3\,\mathbf{i} - 2\,\mathbf{j}) \\ &= 3(1 + \lambda)\,\mathbf{i} + 2(2 - \lambda)\,\mathbf{j}. \end{aligned}$$

The position vectors of each point on the line can be determined for different values of λ as follows:

For $\lambda = 0$, $\mathbf{r} = 3\,\mathbf{i} + 4\,\mathbf{j}$, corresponding to the point A(3, 4) on the line.
For $\lambda = 1$, $\mathbf{r} = 6\,\mathbf{i} + 2\,\mathbf{j}$, corresponding to the point B(6, 2) on the line.
For $\lambda = 2$, $\mathbf{r} = 9\,\mathbf{i}$, corresponding to the point C(9, 0) on the line.

Note that the Cartesian form of the equation of the straight line can also be written as

$$\frac{x - a_x}{b_x} = \frac{y - a_y}{b_y} \quad (= \lambda),$$

where x and y are variables, a_x and a_y are the coordinates of a specific point on the line and the constants b_x and b_y are the x and y

components of the direction vector **b**, respectively.

- The vector equation of a straight line that passes through two given points A and B, as illustrated in figure 1.21, can be expressed as $\mathbf{r} = \mathbf{a} + \lambda(\mathbf{b} - \mathbf{a})$, where **r** is the position vector of any general point P on the straight line with respect to the origin at point O. The position vectors of points A and B with respect to the origin are given by the vectors **a** and **b**, respectively, and λ is a nonzero scalar multiplier that is equal to the ratio AP/AB.

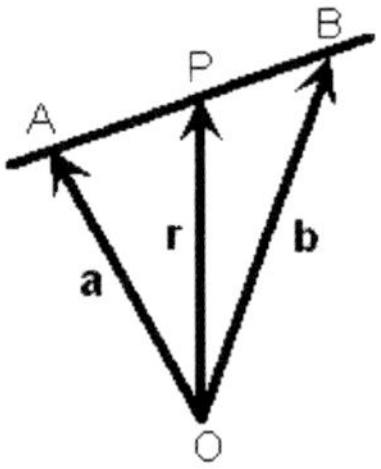

Fig. 1.21. The vector equation $\mathbf{r} = \mathbf{a} + \lambda(\mathbf{b} - \mathbf{a})$ of a straight line that passes through the point with position vector **a** and is parallel to the vector $\mathbf{b} - \mathbf{a}$.

- The condition that three points, A, B and C, which are defined by the position vectors **a**, **b** and **c**, respectively, are collinear is satisfied by the expression

$$\mathbf{a} = \mathbf{b} + \lambda(\mathbf{c} - \mathbf{b}) \text{ or } \mathbf{a} + (\lambda - 1)\mathbf{b} - \lambda\mathbf{c} = \mathbf{0},$$

where λ is a scalar parameter.

- The vector equation of a plane can be determined by either of the following methods.

  1. From the position vectors of three noncollinear points and is given by the expression

  $$\mathbf{r} = \mathbf{a} + \lambda(\mathbf{b} - \mathbf{a}) + \mu(\mathbf{c} - \mathbf{a}),$$

  where **r** is the position vector of any general point P in the plane, and **a**, **b** and **c** are the corresponding position vectors of three known noncollinear points A, B and C that determine the plane, as shown in figure 1.22. Hence, from the vector equation of the plane, you can determine the position vector of one point in the plane given a pair of values for the scalar parameters λ and μ. In the case where the plane passes through the origin at point O, the vector equation of the plane is given by $\mathbf{r} = \lambda(\mathbf{b} - \mathbf{a}) + \mu(\mathbf{c} - \mathbf{a})$.

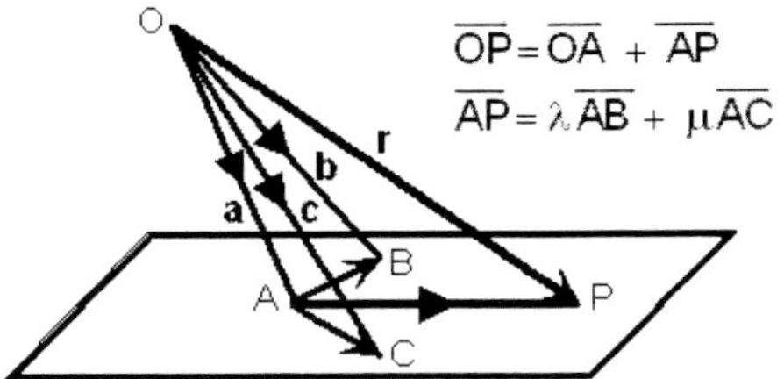

Fig. 1.22. Vector equation of a plane $\mathbf{r} = \mathbf{a} + \lambda(\mathbf{b} - \mathbf{a}) + \mu(\mathbf{c} - \mathbf{a})$, where one and only one plane passes through three noncollinear points, A, B and C.

2. From the intersection of two nonparallel straight lines whose direction vectors are $\mathbf{b}_1$ and $\mathbf{b}_2$, as shown in figure 1.23, and is given by the expression

$$\mathbf{r} = \mathbf{a} + \lambda \mathbf{b}_1 + \mu \mathbf{b}_2,$$

where $\mathbf{r}$ is the position vector of any general point P in the plane and $\mathbf{a}$ is the position vector of the common point A of intersection of the two lines. Hence, from the vector equation of the plane, you can determine the position vector of one point in the plane given a pair of values for the scalar parameters λ and μ.

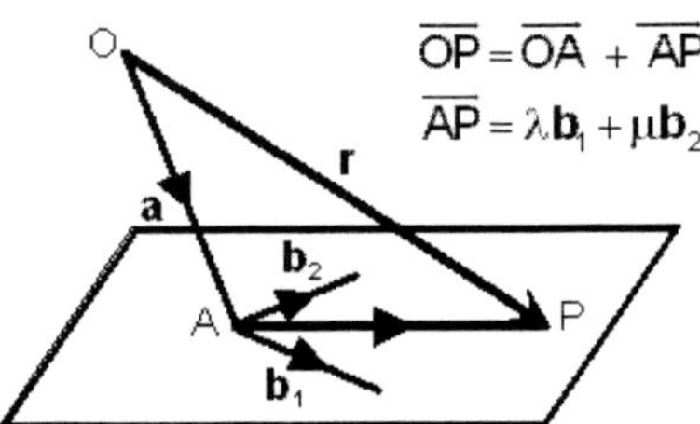

Fig. 1.23. Vector equation of a plane $\mathbf{r} = \mathbf{a} + \lambda \mathbf{b}_1 + \mu \mathbf{b}_2$, where one and only one plane passes through the two intersecting nonparallel straight lines.

- The position vector $\mathbf{g}$ of the centroid of the triangle ABC, i.e., the point of intersection of the triangle's three medians, whose vertices are denoted by the position vectors $\mathbf{a}$, $\mathbf{b}$ and $\mathbf{c}$, is given by

$$\mathbf{g} = \tfrac{1}{3}(\mathbf{a} + \mathbf{b} + \mathbf{c}).$$

1.7 WORKED EXAMPLES

The **Vector Arithmetic Tool** can be used to determine the addition or subtraction of any two (or more) vectors in component form in a Cartesian coordinate system. In addition, the tool can be used to perform the scalar multiplication of vectors by nonzero real numbers. The Vector Arithmetic Tool can also be used to calculate the direction angles and the direction cosines of a vector. See Appendix B for information on how to enter values into any of the Vector Algebra Tools.

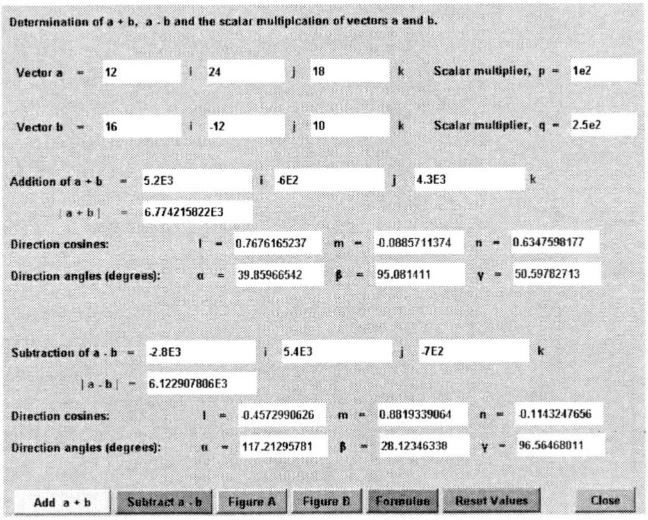

Fig. 1.24. The Vector Arithmetic Tool from the Vector Algebra Tools Software Program.

Question 1. Given the forces $\mathbf{F}_1 = 9\,\mathbf{i} - 14\,\mathbf{j} - 10\,\mathbf{k}$ N and $\mathbf{F}_2 = 12\,\mathbf{i} + 6\,\mathbf{j} + 12\,\mathbf{k}$ N, find the magnitude and direction cosines of (a) $9\mathbf{F}_1 + 3\mathbf{F}_2$ and (b) $3\mathbf{F}_1 - 9\mathbf{F}_2$.

1(a) **Worked Solutions**
Determine the vector addition of $9\mathbf{F}_1 + 3\mathbf{F}_2$, where

$$9\mathbf{F}_1 + 3\mathbf{F}_2 = 9(9\,\mathbf{i} - 14\,\mathbf{j} - 10\,\mathbf{k}) + 3(12\,\mathbf{i} + 6\,\mathbf{j} + 12\,\mathbf{k})$$

$$= 117\,\mathbf{i} - 108\,\mathbf{j} - 54\,\mathbf{k}\ \text{N}.$$

Solve for the magnitude of $9\mathbf{F}_1 + 3\mathbf{F}_2$ given by

$$|\,9\mathbf{F}_1 + 3\mathbf{F}_2\,| = \sqrt{117^2 + (-108)^2 + (-54)^2} = 168.134 \text{ N.}$$

Solve for the direction cosines of $9\mathbf{F}_1 + 3\mathbf{F}_2$ given by

$$\ell = \cos\alpha = \frac{9F_{1x} + 3F_{2x}}{|\,9\,\mathbf{F}_1 + 3\,\mathbf{F}_2\,|} = \frac{81 + 36}{168.134} = 0.6959,$$

$$m = \cos\beta = \frac{9F_{1y} + 3F_{2y}}{|\,9\,\mathbf{F}_1 + 3\,\mathbf{F}_2\,|} = \frac{-126 + 18}{168.134} = -0.6423 \text{ and}$$

$$n = \cos\gamma = \frac{9F_{1z} + 3F_{2z}}{|\,9\,\mathbf{F}_1 + 3\,\mathbf{F}_2\,|} = \frac{-90 + 36}{168.134} = -0.3212.$$

Vector Algebra Tools Solutions

Using the Vector Arithmetic Tool: Enter the given components of vector $\mathbf{F}_1$ in the Vector **a** user input fields and the given components of vector $\mathbf{F}_2$ in the Vector **b** user input fields, together with their respective scalar multipliers in the Scalar multiplier p and q user input fields. Click the Add **a** + **b** button. The answers will be displayed in the output fields as shown.

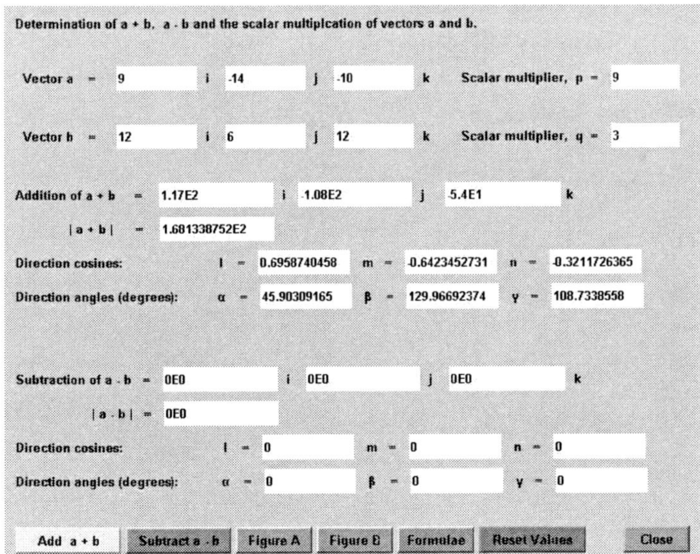

Answers: 1(a) $|\,9\mathbf{F}_1 + 3\mathbf{F}_2\,| = 168.134$ N, $\ell = 0.6959$, $m = -0.6423$ and $n = -0.3212$.

1(b) **Worked Solutions**

Determine the vector subtraction of $3\mathbf{F}_1 - 9\mathbf{F}_2$, where

$$3\mathbf{F}_1 - 9\mathbf{F}_2 = 3(9\,\mathbf{i} - 14\,\mathbf{j} - 10\,\mathbf{k}) - 9(12\,\mathbf{i} + 6\,\mathbf{j} + 12\,\mathbf{k})$$

$$= -81\,\mathbf{i} - 96\,\mathbf{j} - 138\,\mathbf{k}\text{ N.}$$

Solve for the magnitude of $3\mathbf{F}_1 - 9\mathbf{F}_2$ given by

$$|\,3\mathbf{F}_1 - 9\mathbf{F}_2\,| = \sqrt{(-81)^2 + (-96)^2 + (-138)^2} = 186.604\text{ N}$$

Solve for the direction cosines of $3\mathbf{F}_1 - 9\mathbf{F}_2$ given by

$$\ell = \cos\alpha = \frac{3F_{1x} - 9F_{2x}}{|\,3\mathbf{F}_1 - 9\mathbf{F}_2\,|} = \frac{27 - 108}{186.604} = -0.4341,$$

$$m = \cos\beta = \frac{3F_{1y} - 9F_{2y}}{|\,3\mathbf{F}_1 - 9\mathbf{F}_2\,|} = \frac{-42 - 54}{186.604} = -0.5145 \text{ and}$$

$$n = \cos\gamma = \frac{3F_{1z} - 9F_{2z}}{|\,3\mathbf{F}_1 - 9\mathbf{F}_2\,|} = \frac{-30 - 108}{186.604} = -0.7395.$$

Vector Algebra Tools Solutions

Using the Vector Arithmetic Tool: Enter the given components of vector $\mathbf{F}_1$ in the Vector **a** user input fields and the given components of vector $\mathbf{F}_2$ in the Vector **b** user input fields, together with their respective scalar multipliers in the Scalar multiplier p and q user input fields. Click the Subtract **a** − **b** button. The answers will be displayed in the output fields as shown.

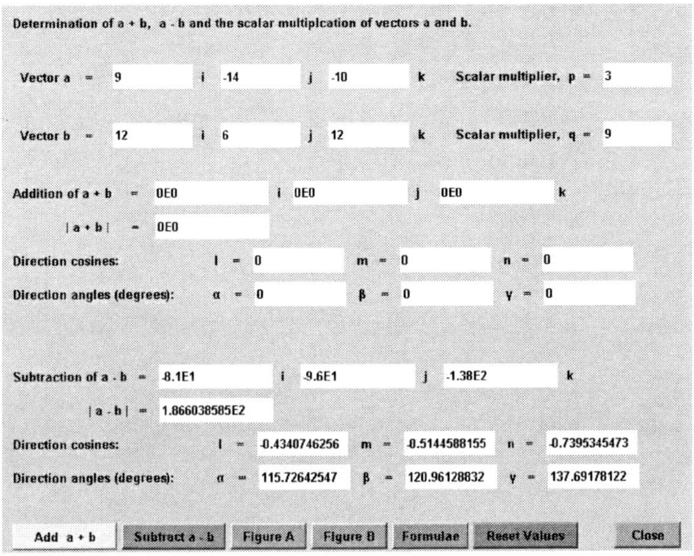

Determination of a + b, a - b and the scalar multiplcation of vectors a and b.

Vector a =	9	i	-14	j	-10	k	Scalar multiplier, p = 3
Vector b =	12	i	6	j	12	k	Scalar multiplier, q = 9

Addition of a + b =	0E0	i	0E0	j	0E0	k
| a + b | =	0E0					
Direction cosines:	l =	0	m =	0	n =	0
Direction angles (degrees):	α =	0	β =	0	γ =	0
Subtraction of a - b =	-8.1E1	i	-9.6E1	j	-1.38E2	k
| a - b | =	1.866038585E2					
Direction cosines:	l =	-0.4340746256	m =	-0.5144588155	n =	-0.7395345473
Direction angles (degrees):	α =	115.72642547	β =	120.96128832	γ =	137.69178122

Add a + b | Subtract a - b | Figure A | Figure B | Formulae | Reset Values | Close

Answers: 1(b) $|3\mathbf{F}_1 - 9\mathbf{F}_2| = 186.604$ N, $\ell = -0.4341$, $m = -0.5145$ and $n = -0.7395$.

Question 2. The position vector $\overrightarrow{OP}$ of the point P from the origin O that divides the line AB into the ratio m:n, such that AP/PB = m/n, as shown in figure 1.25, is given by the expression

$$\overrightarrow{OP} = \frac{n\mathbf{a} + m\mathbf{b}}{m + n}.$$

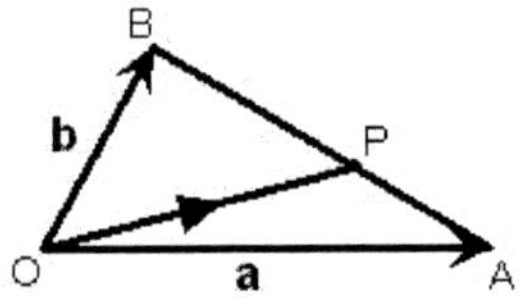

Fig. 1.25.

Determine (a) the vector $\overrightarrow{OP}$ and (b) the length of $\overrightarrow{OP}$ if the line AB is divided into the ratio 3:5, where the points A and B have position vectors $\mathbf{a} = 2\,\mathbf{i} - 7\,\mathbf{j} + 4\,\mathbf{k}$ m and $\mathbf{b} = 4\,\mathbf{i} + 3\,\mathbf{j} + 6\,\mathbf{k}$ m, respectively.

Worked Solutions

Determine the vector addition of $n\mathbf{a} + m\mathbf{b}$, where

$$n\mathbf{a} + m\mathbf{b} = 5(2\,\mathbf{i} - 7\,\mathbf{j} + 4\,\mathbf{k}) + 3(4\,\mathbf{i} + 3\,\mathbf{j} + 6\,\mathbf{k})$$

$$5\mathbf{a} + 3\mathbf{b} = 22\ \mathbf{i} - 26\ \mathbf{j} + 38\ \mathbf{k}\ \text{m}.$$

Solve for vector $\overrightarrow{OP}$, where

$$\overrightarrow{OP} = 0.125(22\ \mathbf{i} - 26\ \mathbf{j} + 38\ \mathbf{k}) = 2.75\ \mathbf{i} - 3.25\ \mathbf{j} + 4.75\ \mathbf{k}\ \text{m}.$$

The length of $\overrightarrow{OP}$, is given by

$$|\overrightarrow{OP}| = \sqrt{2.75^2 + (-3.25)^2 + 4.75^2} = 6.3787\ \text{m}.$$

Vector Algebra Tools Solutions

Using the Vector Arithmetic Tool: Enter the given components of vector **a** in the Vector **a** user input fields and the given components of vector **b** in the Vector **b** user input fields, together with their respective scalar multipliers in the Scalar multiplier p and q user input fields. Then click the Add **a** + **b** button. The answers will be displayed in the output fields as shown. Click the Reset Values button before continuing to the next step.

Determination of a + b, a - b and the scalar multiplcation of vectors a and b.

Vector a	=	2	i	-7	j	4	k	Scalar multiplier, p =	5
Vector b	=	4	i	3	j	6	k	Scalar multiplier, q =	3
Addition of a + b	=	2.2E1	i	-2.6E1	j	3.8E1	k		
|a + b|	=	5.102940329E1							
Direction cosines:	l =	0.4311239909	m =	-0.5095101711	n =	0.7446687116			
Direction angles (degrees):	α =	64.4610874	β =	120.63120799	γ =	41.86934835			
Subtraction of a - b	=	0E0	i	0E0	j	0E0	k		
|a - b|	=	0E0							
Direction cosines:	l =	0	m =	0	n =	0			
Direction angles (degrees):	α =	0	β =	0	γ =	0			

Add a + b | Subtract a - b | Figure A | Figure B | Formulae | Reset Values | Close

Solution: $n\mathbf{a} + m\mathbf{b} = 5\mathbf{a} + 3\mathbf{b} = 22\ \mathbf{i} - 26\ \mathbf{j} + 38\ \mathbf{k}$ m.

Using the Vector Arithmetic Tool: Enter the calculated components of the resultant vector $n\mathbf{a} + m\mathbf{b}$ in the Vector **a** user input fields, together with a value of 0.125 in the Scalar multiplier p user input field. Enter 0 in each of the Vector **b** user input fields. Then click the Add **a** + **b** button. The answers will be displayed in the output fields as shown.

Determination of a + b, a - b and the scalar multiplcation of vectors a and b.

Vector a =	22	i	-26	j	38	k	Scalar multiplier, p =	0.125
Vector b =	0	i	0	j	0	k	Scalar multiplier, q =	1

Addition of a + b =	2.75E0	i	-3.25E0	j	4.75E0	k
|a + b| =	6.378675411E0					
Direction cosines:	l =	0.4311239909	m =	-0.5095101711	n =	0.7446687116
Direction angles (degrees):	α =	64.4610874	β =	120.63120799	γ =	41.86934835

Subtraction of a - b =	0E0	i	0E0	j	0E0	k
|a - b| =	0E0					
Direction cosines:	l =	0	m =	0	n =	0
Direction angles (degrees):	α =	0	β =	0	γ =	0

Add a + b | Subtract a - b | Figure A | Figure B | Formulae | Reset Values | Close

Solutions: 2(a) $\overrightarrow{OP} = 2.75\,\mathbf{i} - 3.25\,\mathbf{j} + 4.75\,\mathbf{k}$ m and (b) $|\overrightarrow{OP}| = 6.3787$ m.

Answers: 2(a) $\overrightarrow{OP} = 2.75\,\mathbf{i} - 3.25\,\mathbf{j} + 4.75\,\mathbf{k}$ m and (b) $|\overrightarrow{OP}| = 6.3787$ m. Note that in the case where $m = n = 1$, then $\overrightarrow{OP}$ is the position vector of the point P that is the mid-point of the line AB and is given by ½ (**a** + **b**).

Question 3. Two force vectors defined by $\mathbf{F}_1 = 16\,\mathbf{i} + 16\,\mathbf{j} - 12\,\mathbf{k}$ N and $\mathbf{F}_2 = 12\,\mathbf{i} + 20\,\mathbf{j} - 12\,\mathbf{k}$ N act on a mass of 20 kg located at the origin of a three-dimensional coordinate system. Determine (a) the acceleration vector **a**, (b) the magnitude of the acceleration vector **a** and (c) the angles that the acceleration vector **a** makes with the positive *x*, *y* and *z*-axes.

Worked Solutions

The acceleration vector **a** of the mass *m* is given by Newton's second law of motion where the net external force acting on *m* is defined by the equation $\mathbf{F}_{\text{net}} = \sum\mathbf{F} = m\mathbf{a}$. The angles between the acceleration vector **a** and the positive *x*, *y* and *z*-axes are given by the direction angles, which can be determined from the direction cosines of **a**.

(a) The net external force $\mathbf{F}_{net}$ acting on the mass m is

$$\mathbf{F}_{net} = \sum \mathbf{F} = \mathbf{F}_1 + \mathbf{F}_2$$

$$= (16\,\mathbf{i} + 16\,\mathbf{j} - 12\,\mathbf{k}) + (12\,\mathbf{i} + 20\,\mathbf{j} - 12\,\mathbf{k})$$

$$= 28\,\mathbf{i} + 36\,\mathbf{j} - 24\,\mathbf{k}\text{ N.}$$

From Newton's second law of motion, the acceleration vector **a** is given by

$$\mathbf{a} = \frac{\mathbf{F}_{net}}{m} = \frac{1}{20}(28\,\mathbf{i} + 36\,\mathbf{j} - 24\,\mathbf{k}) = 1.4\,\mathbf{i} + 1.8\,\mathbf{j} - 1.2\,\mathbf{k}\text{ m/s}^2.$$

(b) The magnitude of **a**, | **a** |, is

$$|\,\mathbf{a}\,| = \sqrt{1.4^2 + 1.8^2 + (-1.2)^2} = 2.5768\text{ m/s}^2.$$

(c) The angles α, β and γ that the acceleration vector **a** makes with the positive x, y and z-axes, respectively, are the direction angles, and are determined as follows:

$$\cos\alpha = \frac{1.4}{2.5768} = 0.5433,\text{ where } \alpha = \cos^{-1}(0.5433) = 57.09°,$$

$$\cos\beta = \frac{1.8}{2.5768} = 0.6985,\text{ where } \beta = \cos^{-1}(0.6985) = 45.69° \text{ and}$$

$$\cos\gamma = \frac{-1.2}{2.5768} = -0.4657,\text{ where } \gamma = \cos^{-1}(-0.4657) = 117.76°.$$

Vector Algebra Tools Solutions

Using the Vector Arithmetic Tool: Enter the given components of vector $\mathbf{F}_1$ in the Vector **a** user input fields and the given components of vector $\mathbf{F}_2$ in the Vector **b** user input fields, together with a value of 0.05 in both the Scalar multiplier p and the Scalar multiplier q user input fields. Then click the Add **a** + **b** button. The answers will be displayed in the output fields as shown.

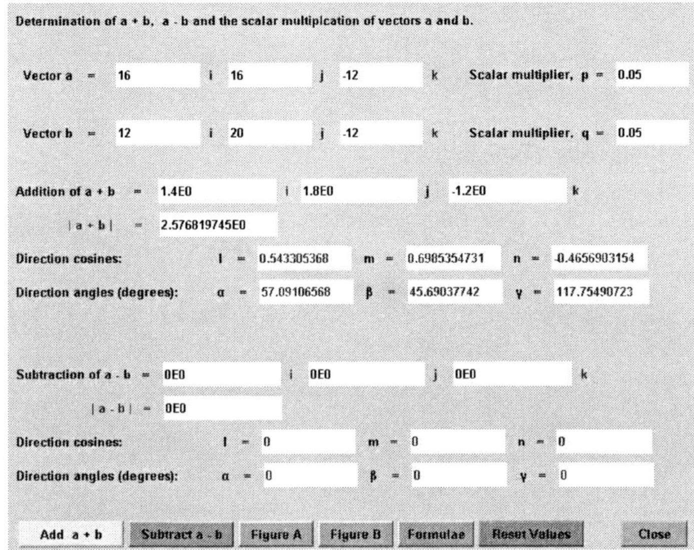

Answers: 3(a) **a** = 1.4 **i** + 1.8 **j** - 1.2 **k** m/s^2, 3(b) | **a** | = 2.5768 m/s^2 and 3(c) α = 57.09°, β = 45.69° and γ = 117.76°.

Question 4. Determine the magnitude of the net force on a positive charge q of 3.5 μC in the presence of two electric fields given by $\mathbf{E}_1 = 10^4\,(6\,\mathbf{i} + 1.5\,\mathbf{j} - 2.7\,\mathbf{k})$ N/C and $\mathbf{E}_2 = 2.3 \times 10^4\,\mathbf{i} - 4.5 \times 10^5\,\mathbf{j} + 8 \times 10^4\,\mathbf{k}$ N/C.

Worked Solution

First solve for the net force, $\mathbf{F}_{\text{net}}$, corresponding to the superposition of forces, i.e., the vector sum of the forces due to the two electric fields $\mathbf{E}_1$ and $\mathbf{E}_2$ acting on the positive charge q, where

$$\mathbf{F}_{\text{net}} = q\mathbf{E}_1 + q\mathbf{E}_2$$

$$= q(\mathbf{E}_1 + \mathbf{E}_2)$$

$$= 3.5 \times 10^{-6}\,[(6 \times 10^4\,\mathbf{i} + 1.5 \times 10^4\,\mathbf{j} - 2.7 \times 10^4\,\mathbf{k}) + (2.3 \times 10^4\,\mathbf{i} - 4.5 \times 10^5\,\mathbf{j} + 8 \times 10^4\,\mathbf{k})]$$

$$= 3.5 \times 10^{-6}\,(8.3 \times 10^4\,\mathbf{i} - 4.35 \times 10^5\,\mathbf{j} + 5.3 \times 10^4\,\mathbf{k})$$

$$= 0.2905\,\mathbf{i} - 1.5225\,\mathbf{j} + 0.1855\,\mathbf{k}\ \text{N}.$$

Then determine the magnitude of $\mathbf{F}_{net}$, $|\mathbf{F}_{net}|$, where

$$|\mathbf{F}_{net}| = \sqrt{0.2905^2 + (-1.5225)^2 + 0.1855^2} = 1.561 \text{ N}.$$

Vector Algebra Tools Solution

Using the Vector Arithmetic Tool: Enter the given components of vector $\mathbf{E}_1$ in the Vector **a** user input fields and the given components of vector $\mathbf{E}_2$ in the Vector **b** user input fields, together with a value of 3.5×10^{-6} for the positive charge q in both the Scalar multiplier p and q user input fields. Then click the Add **a** + **b** button. The answers will be displayed in the output fields as shown.

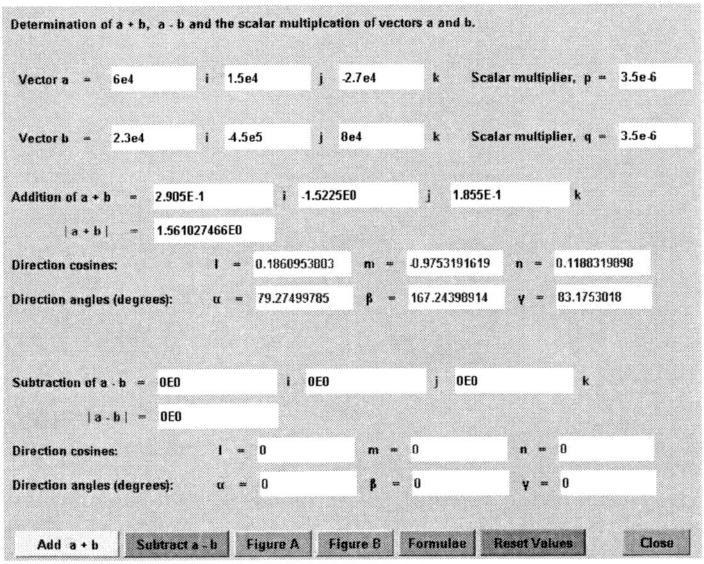

Determination of a + b, a - b and the scalar multiplcation of vectors a and b.

Vector a = 6e4 i 1.5e4 j -2.7e4 k Scalar multiplier, p = 3.5e-6

Vector b = 2.3e4 i -4.5e5 j 8e4 k Scalar multiplier, q = 3.5e-6

Addition of a + b = 2.905E-1 i -1.5225E0 j 1.855E-1 k

| a + b | = 1.561027466E0

Direction cosines: l = 0.1860953803 m = -0.9753191619 n = 0.1188319898

Direction angles (degrees): α = 79.27499785 β = 167.24398914 γ = 83.1753018

Subtraction of a - b = 0E0 i 0E0 j 0E0 k

| a - b | = 0E0

Direction cosines: l = 0 m = 0 n = 0

Direction angles (degrees): α = 0 β = 0 γ = 0

Add a + b | Subtract a - b | Figure A | Figure B | Formulae | Reset Values | Close

Solutions: $\mathbf{F}_{net} = 0.2905\,\mathbf{i} - 1.5225\,\mathbf{j} + 0.1855\,\mathbf{k}$ N and $|\mathbf{F}_{net}| = 1.561$ N.

Answer: The magnitude of the net force on the electric charge q is $|\mathbf{F}_{net}| = 1.561$ N.

1.8 NOTES:

CHAPTER 2

VECTOR COMPOSITION

In this chapter the unit vector and the synthesis of a vector are presented. Definitions, problem solving applications and numerous worked examples of unit vector determinations and vector composition are included. Also demonstrated is the use of the Vector Composition software tool in solving the example problems.

2.1 DEFINITION AND COMPOSITION OF A VECTOR

A vector is a physical quantity that has both magnitude and direction. Therefore, any nonzero vector **a** can be expressed as the product of its magnitude and its unit vector, such that

$$\mathbf{a} = |\,\mathbf{a}\,|\,\hat{\mathbf{a}},$$

where $|\,\mathbf{a}\,|$ is the magnitude of vector **a** and $\hat{\mathbf{a}}$ is the unit vector that is in the direction of vector **a**. Consequently, if vector **a** is expressed in component form as $\mathbf{a} = a_x\,\mathbf{i} + a_y\,\mathbf{j} + a_z\,\mathbf{k}$, then

$$\mathbf{a} = |\,\mathbf{a}\,|\,\hat{\mathbf{a}} = |\,\mathbf{a}\,|\frac{a_x\,\mathbf{i} + a_y\,\mathbf{j} + a_z\,\mathbf{k}}{\sqrt{a_x^{\,2} + a_y^{\,2} + a_z^{\,2}}}.$$

Vector **a** also can be expressed as the product of its magnitude and the unit vector $\hat{\mathbf{n}}$ of any collinear vector **n** that is in the same direction as **a**, where $\mathbf{n} = n_x\,\mathbf{i} + n_y\,\mathbf{j} + n_z\,\mathbf{k}$, and is given by

$$\mathbf{a} = |\,\mathbf{a}\,|\,\hat{\mathbf{n}} = |\,\mathbf{a}\,|\,\frac{\mathbf{n}}{|\,\mathbf{n}\,|} = |\,\mathbf{a}\,|\,\frac{n_x\,\mathbf{i} + n_y\,\mathbf{j} + n_z\,\mathbf{k}}{\sqrt{n_x^{\,2} + n_y^{\,2} + n_z^{\,2}}}.$$

2.2 DEFINITION AND COMPOSITION OF A UNIT VECTOR

A **unit vector**, sometimes referred to as a **normalized vector**, is a dimensionless vector that has a magnitude of one and a specific direction. In print unit vectors are distinguished from other vectors by the use of a circumflex. For example, the unit vector $\hat{\mathbf{a}}$ which specifies the direction of vector **a** is given by the expression

$$\hat{\mathbf{a}} = \frac{\mathbf{a}}{|\,\mathbf{a}\,|},$$

where $|\,\mathbf{a}\,|$ is the magnitude of the nonzero vector **a**. The magnitude or length of the unit vector $\hat{\mathbf{a}}$ equals one. The process of creating a unit vector from a vector of arbitrary length is known as **normalizing** a vector. A graphical representation of vector **a** and its corresponding unit vector $\hat{\mathbf{a}}$ is shown in the figure 2.1.

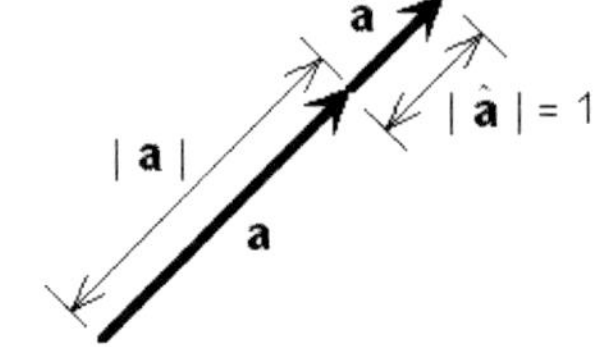

Fig. 2.1. Vector **a** is expressed as $\mathbf{a} = |\,\mathbf{a}\,|\ \hat{\mathbf{a}}$, where $\hat{\mathbf{a}}$ is a unit vector that has a magnitude of 1 and is in the same direction as **a**.

Shown in figure 2.2 are the Cartesian axes of a three-dimensional coordinate system where the base vectors are the three mutually orthogonal unit vectors **i**, **j** and **k** that are used to specify the directions of the positive *x*, *y* and *z*-axes, respectively. The corresponding magnitudes of the **i**, **j** and **k** unit vectors are equal to one and can be written as

$$|\,\mathbf{i}\,| = |\,\mathbf{j}\,| = |\,\mathbf{k}\,| = 1.$$

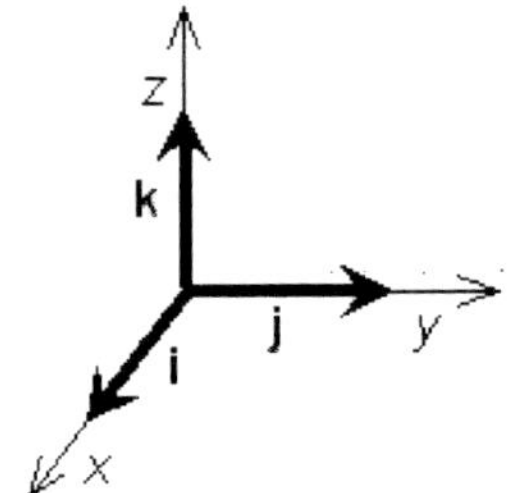

Fig. 2.2. The three mutually orthogonal unit vectors **i**, **j** and **k** are directed parallel to the positive *x*, *y* and *z*-axes, respectively.

If vector **a** is represented in component form as $\mathbf{a} = a_x\,\mathbf{i} + a_y\,\mathbf{j} + a_z\,\mathbf{k}$, then the unit vector $\hat{\mathbf{a}}$ that is in the same direction as vector **a** can be expressed as

$$\hat{\mathbf{a}} = \frac{\mathbf{a}}{|\,\mathbf{a}\,|} = \frac{a_x\,\mathbf{i} + a_y\,\mathbf{j} + a_z\,\mathbf{k}}{\sqrt{a_x^{\,2} + a_y^{\,2} + a_z^{\,2}}},$$

where the magnitude of **a** is given by $|\,\mathbf{a}\,| = \sqrt{a_x^{\,2} + a_y^{\,2} + a_z^{\,2}}$.

Alternatively, the unit vector $\hat{\mathbf{a}}$ can also be expressed as

$$\hat{\mathbf{a}} = \frac{a_x}{|\mathbf{a}|}\mathbf{i} + \frac{a_y}{|\mathbf{a}|}\mathbf{j} + \frac{a_z}{|\mathbf{a}|}\mathbf{k},$$

where **components** of $\hat{\mathbf{a}}$ are the **direction cosines** $[\ell, m, n]$ of vector **a** and are defined as follows:

$$\ell = \cos\alpha = \frac{a_x}{|\mathbf{a}|}, \ m = \cos\beta = \frac{a_y}{|\mathbf{a}|} \text{ and } n = \cos\gamma = \frac{a_z}{|\mathbf{a}|},$$

such that

$$\hat{\mathbf{a}} \bullet \mathbf{i} = \ell, \ \hat{\mathbf{a}} \bullet \mathbf{j} = m \text{ and } \hat{\mathbf{a}} \bullet \mathbf{k} = n.$$

Recall that the angles α, β and γ are the **direction angles** that vector **a** makes with the positive x, y, and z-axes. The direction cosines are not independent, as the sum of the squares of the direction cosines for any vector must satisfy the following condition:

$$\ell^2 + m^2 + n^2 = \cos^2\alpha + \cos^2\beta + \cos^2\gamma = \frac{a_x^{\ 2} + a_y^{\ 2} + a_z^{\ 2}}{|\mathbf{a}|^2} = 1.$$

Since, the direction cosines of vector **a** are the **i**, **j** and **k** components of $\hat{\mathbf{a}}$, we can write

$$\hat{\mathbf{a}} = \ell\mathbf{i} + m\,\mathbf{j} + n\,\mathbf{k} = \cos\alpha\,\mathbf{i} + \cos\beta\,\mathbf{j} + \cos\gamma\,\mathbf{k},$$

where $|\hat{\mathbf{a}}| = \sqrt{\ell^2 + m^2 + n^2} = 1.$

Conversely, if

$$\cos(180° - \alpha) = -\ell, \ \cos(180° - \beta) = -m \text{ and } \cos(180° - \gamma) = -n,$$

then the corresponding unit vector of vector **a** can be expressed as

$$\hat{\mathbf{a}} = -\ell\mathbf{i} - m\,\mathbf{j} - n\,\mathbf{k},$$

indicating that vector **a** now points in the opposite direction.

For the collinear vectors **a** and **b**, where $\mathbf{a} = \lambda\mathbf{b}$, both **a** and **b** point in the same direction and have direction cosines that are equal in magnitude and sign when λ is positive. However, when λ is negative, both **a** and **b** point in opposite directions and have direction cosines that are equal in magnitude but opposite in sign.

In summary, any vector **a** can be defined as the product of its magnitude and a unit vector that is in the direction of vector **a**. Specifically, the components of the unit vector are the vector's direction cosines, which uniquely determine the vector's direction. Therefore, any vector **a** can be expressed as

$$\mathbf{a} = |\,\mathbf{a}\,|\,\hat{\mathbf{a}} = |\,\mathbf{a}\,|\,(\ell\,\mathbf{i} + m\,\mathbf{j} + n\,\mathbf{k}) = |\,\mathbf{a}\,|\,(\cos\alpha\,\mathbf{i} + \cos\beta\,\mathbf{j} + \cos\gamma\,\mathbf{k}).$$

Note that, for any collinear vector **n** that points in the same direction as vector **a**, the components of unit vector $\hat{\mathbf{n}}$ are also the direction cosines as vector **a**. Hence, since $\hat{\mathbf{n}}$ and $\hat{\mathbf{a}}$ are equal, unit vector $\hat{\mathbf{n}}$ can be used to determine vector **a** when both $\hat{\mathbf{n}}$ and the magnitude of vector **a** are known. This method is commonly used to determine a vector when a collinear vector is used to specify the direction of a vector of known magnitude. For an example, see Question 2 in the Worked Examples section of this Chapter.

2.3 DETERMINATION OF VECTOR QUANTITIES

2.3.1 Synthesis of a vector

The following vector composition methods are commonly used to determine vector **a** from its magnitude and direction:

1. Vector **a** can be expressed as the product of its magnitude, $|\,\mathbf{a}\,|$, and its unit vector $\hat{\mathbf{a}}$ that is in the direction of vector **a**, such that

$$\mathbf{a} = |\,\mathbf{a}\,|\,\hat{\mathbf{a}}\;.$$

2. Vector **a** can be expressed as the product of its magnitude and the unit vector $\hat{\mathbf{n}}$ of any collinear vector **n** that is in the same direction as vector **a**, where $\mathbf{n} = n_x\,\mathbf{i} + n_y\,\mathbf{j} + n_z\,\mathbf{k}$, such that

$$\mathbf{a} = |\,\mathbf{a}\,|\,\hat{\mathbf{n}} = |\,\mathbf{a}\,|\frac{\mathbf{n}}{|\,\mathbf{n}\,|} = |\,\mathbf{a}\,|\,\frac{n_x\,\mathbf{i} + n_y\,\mathbf{j} + n_z\,\mathbf{k}}{\sqrt{n_x^{\,2} + n_y^{\,2} + n_z^{\,2}}}\,.$$

Note that the collinear vector **n** that is used to specify the direction of a given vector is also referred to as the **direction vector**.

3. Vector **a** can be expressed as the product of its magnitude and its unit vector $\hat{\mathbf{a}}$ or the unit vector $\hat{\mathbf{n}}$ of any collinear vector **n** that is in the same direction as vector **a**, where the components of $\hat{\mathbf{a}}$ and $\hat{\mathbf{n}}$ are the direction cosines of both vectors, such that

$$\mathbf{a} = |\,\mathbf{a}\,|\,\hat{\mathbf{a}} = |\,\mathbf{a}\,|\,\hat{\mathbf{n}} = |\,\mathbf{a}\,|\,(\ell\,\mathbf{i} + m\,\mathbf{j} + n\,\mathbf{k})$$

$$= |\,\mathbf{a}\,|\,(\cos\alpha\,\mathbf{i} + \cos\beta\,\mathbf{j} + \cos\gamma\,\mathbf{k}),$$

where [ℓ, m, n] are the direction cosines of vectors **a** or **n**.

2.3.2 Determination of a unit vector

The unit vector $\hat{\mathbf{a}}$ that is in the direction of vector **a**, where $\mathbf{a} = a_x\,\mathbf{i} + a_y\,\mathbf{j} + a_z\,\mathbf{k}$, can be expressed in component form as

$$\hat{\mathbf{a}} = \frac{\mathbf{a}}{|\,\mathbf{a}\,|} = \frac{a_x\,\mathbf{i} + a_y\,\mathbf{j} + a_z\,\mathbf{k}}{\sqrt{a_x^2 + a_y^2 + a_z^2}}.$$

Alternatively, unit vector $\hat{\mathbf{a}}$ can also be expressed as

$$\hat{\mathbf{a}} = \ell\,\mathbf{i} + m\,\mathbf{j} + n\,\mathbf{k} = \cos\alpha\,\mathbf{i} + \cos\beta\,\mathbf{j} + \cos\gamma\,\mathbf{k},$$

where the direction cosines of vector **a** are the components of $\hat{\mathbf{a}}$.

2.4 PROBLEM SOLVING APPLICATIONS OF VECTOR SYNTHESIS AND UNIT VECTOR DETERMINATIONS

The fundamental methods discussed above demonstrating the synthesis of a vector or unit vector are widely used in solving vector algebra problems. Some of the more common applications of these methods include the following:

- Determination of the force vector **F** that is expressed as the product of its magnitude $|\,\mathbf{F}\,|$ and a unit vector $\hat{\mathbf{f}}$ that is in the direction of **F**, such that

  $$\mathbf{F} = |\,\mathbf{F}\,|\;\hat{\mathbf{f}}.$$

  For example, the force vector **F** of magnitude 30 N, whose unit vector $\left(\frac{2}{3}\mathbf{i} - \frac{2}{3}\mathbf{j} + \frac{1}{3}\mathbf{k}\right)$ is in the direction of **F**, is given by

  $$\mathbf{F} = 30\left(\frac{2}{3}\mathbf{i} - \frac{2}{3}\mathbf{j} + \frac{1}{3}\mathbf{k}\right) = 20\,\mathbf{i} - 20\,\mathbf{j} + 10\,\mathbf{k}\ \text{N}.$$

- Determination of the force vector **F** that is expressed as the product of its magnitude $|\,\mathbf{F}\,|$ and a unit vector $\hat{\mathbf{n}}$ of any collinear vector **n** that is in the same direction as **F**, where $\mathbf{n} = n_x\,\mathbf{i} + n_y\,\mathbf{j} + n_z\,\mathbf{k}$, such that

$$\mathbf{F} = |\,\mathbf{F}\,|\,\hat{\mathbf{n}} = |\,\mathbf{F}\,|\,\frac{\mathbf{n}}{|\,\mathbf{n}\,|} = |\,\mathbf{F}\,|\,\frac{n_x\,\mathbf{i} + n_y\,\mathbf{j} + n_z\,\mathbf{k}}{\sqrt{n_x^{\,2} + n_y^{\,2} + n_z^{\,2}}}.$$

For example, a force vector **F** of magnitude 30 N that is applied in the direction of the vector 2 **i** – 2 **j** + **k**, is given by

$$\mathbf{F} = 30\,\frac{2\,\mathbf{i} - 2\,\mathbf{j} + \mathbf{k}}{\sqrt{2^2 + (-2)^2 + 1^2}} = 20\,\mathbf{i} - 20\,\mathbf{j} + 10\,\mathbf{k}\ \text{N}.$$

- Determination of the force vector **F** that is expressed as the product of its magnitude | **F** | and either its unit vector $\hat{\mathbf{f}}$ or the unit vector $\hat{\mathbf{n}}$ of any collinear vector **n** that is the same direction as **F**, where the components of $\hat{\mathbf{f}}$ and $\hat{\mathbf{n}}$ are the direction cosines of **F**, such that

$$\mathbf{F} = |\,\mathbf{F}\,|\,(\ell\,\mathbf{i} + m\,\mathbf{j} + n\,\mathbf{k}) = |\,\mathbf{F}\,|\,(\cos\alpha\,\mathbf{i} + \cos\beta\,\mathbf{j} + \cos\gamma\,\mathbf{k}).$$

For example, a force vector $\mathbf{F}_1$ of magnitude 30 N that is applied in the direction of the vector $\mathbf{F}_2$ whose direction cosines are $\left[\frac{2}{3}, \frac{-2}{3}, \frac{1}{3}\right]$, is given by

$$\mathbf{F}_1 = 30\left(\frac{2}{3}\mathbf{i} - \frac{2}{3}\mathbf{j} + \frac{1}{3}\mathbf{k}\right) = 20\,\mathbf{i} - 20\,\mathbf{j} + 10\,\mathbf{k}\ \text{N}.$$

- Determination of the unit vector $\hat{\mathbf{n}}$ of any vector **n** that is collinear with vector **a**, such that $\mathbf{n} = \lambda\mathbf{a}$. When $\lambda > 0$, vectors **n** and **a** point in the same direction and are considered parallel. Conversely, when $\lambda < 0$, vectors **n** and **a** point in opposite directions and are anti-parallel. See figure 2.3. Thus, the unit vector $\hat{\mathbf{n}}$ of any vector **n** that is collinear to vector $\lambda\mathbf{a}$ is given by

$$\hat{\mathbf{n}} = \frac{\mathbf{n}}{|\,\mathbf{n}\,|} = \frac{\lambda\mathbf{a}}{|\,\lambda\mathbf{a}\,|}.$$

Clearly, if λ is positive, the direction cosines for vectors **n** and **a** are the same. However, if λ is negative the direction cosines of vector **n** are equal in magnitude but opposite in sign to those of vector **a**. Therefore, parallel vectors have equal direction cosines whereas anti-parallel vectors have direction cosines that are equal in magnitude but opposite in sign.

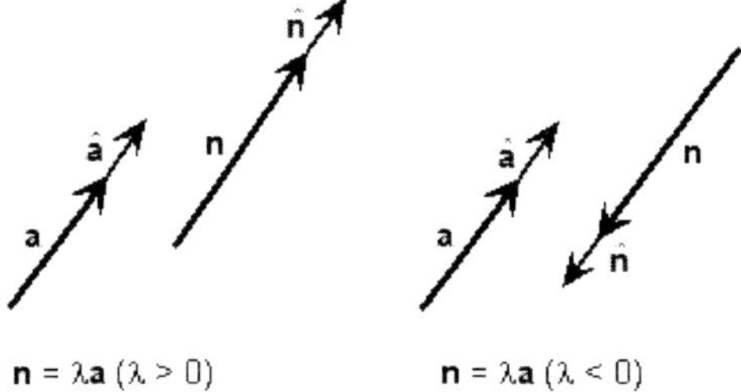

Fig. 2.3. Corresponding unit vectors $\hat{\mathbf{a}}$ and $\hat{\mathbf{n}}$ specifying the directions of the collinear vectors **a** and **n**, respectively, where $\mathbf{n} = \lambda\mathbf{a}$.

- Applications of vector algebra in plane geometry includes the construction of the vector equation of a straight line or a plane and the determination of the position vector of a point on a straight line, of the angle between two straight lines, of the angle between two intersecting planes and of the unit vector normal to a plane.

- The electric flux Φ_E through a surface of fixed area A in a uniform electric field **E** can be expressed as $\Phi_E = \mathbf{E} \bullet \mathbf{A}$, where **A** is the directed area vector normal to the surface, such that $\mathbf{A} = |\,\mathbf{A}\,|\,\hat{\mathbf{n}} = A\,\hat{\mathbf{n}}$. The unit vector $\hat{\mathbf{n}}$ is in the direction of vector **A** and, by convention, points outwards from the surface.

- Similarly, the magnetic flux Φ_B through a surface of fixed area A in a uniform magnetic field **B** can be expressed as $\Phi_B = \mathbf{B} \bullet \mathbf{A}$, where **A** is the directed area vector normal to the surface, such that $\mathbf{A} = |\,\mathbf{A}\,|\,\hat{\mathbf{n}} = A\,\hat{\mathbf{n}}$. The unit vector $\hat{\mathbf{n}}$ is in the direction of vector **A** and, by convention, points outwards from the surface.

2.5 WORKED EXAMPLES

The **Vector Composition Tool** can be used to determine a vector given the vector's magnitude and a collinear vector that specifies its direction in a Cartesian coordinate system. The Vector Composition Tool can also be used to determine a vector given its magnitude and the direction cosines of a vector or a collinear vector that is in the same direction as the vector. In addition, the Vector Composition Tool can be used to determine the unit vector from a vector or from a collinear vector that is in the same direction as the vector. See Appendix B for information on how to enter values into any of the Vector Algebra Tools.

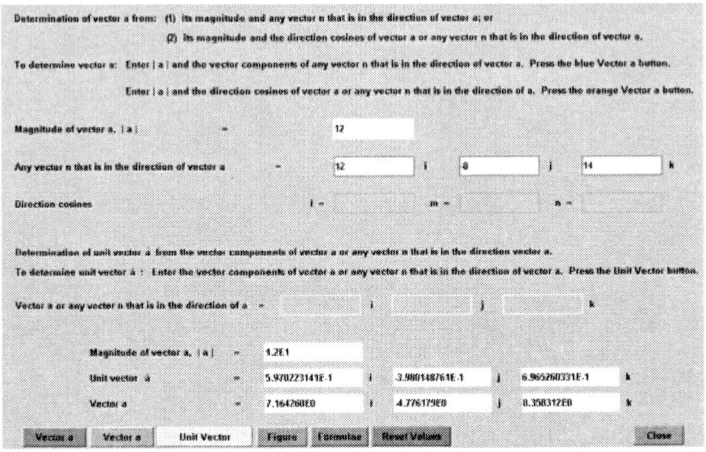

Fig. 2.4. The Vector Composition Tool from the Vector Algebra Tools Software Program.

Question 1. Determine (a) the unit vector $\hat{\mathbf{s}}$ that is in the direction of the displacement vector **s**, where $\mathbf{s} = 3\,\mathbf{i} - 6\,\mathbf{j} + 8\,\mathbf{k}$ m, (b) the direction cosines of the displacement vector **s** and (c) the unit vector of the collinear vector $\mathbf{c} = \lambda\mathbf{s}$, where $\lambda = -5$.

1(a) **Worked Solution**

Solve for $\hat{\mathbf{s}}$, where

$$\hat{\mathbf{s}} = \frac{\mathbf{s}}{|\,\mathbf{s}\,|} = \frac{3\mathbf{i} - 6\,\mathbf{j} + 8\,\mathbf{k}}{\sqrt{3^2 + (-6)^2 + 8^2}} = 0.2873\,\mathbf{i} - 0.5747\,\mathbf{j} + 0.7663\,\mathbf{k}.$$

Vector Algebra Tools Solution

Using the Vector Composition Tool: Enter the given components of vector **s** in the Vector **a** or any vector **n** that is in the direction of **a** user input fields. Click the Unit Vector button. The answers will be displayed in the output fields as shown.

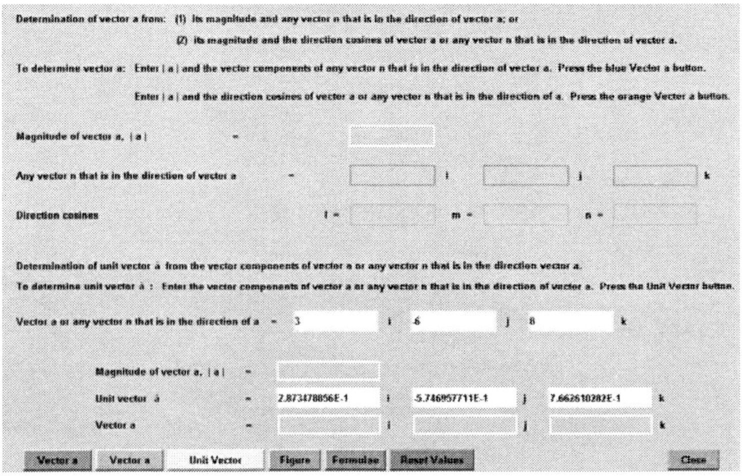

Determination of vector a from: (1) its magnitude and any vector n that is in the direction of vector a; or

(2) its magnitude and the direction cosines of vector a or any vector n that is in the direction of vector a.

To determine vector a: Enter | a | and the vector components of any vector n that is in the direction of vector a. Press the blue Vector a button.

Enter | a | and the direction cosines of vector a or any vector n that is in the direction of a. Press the orange Vector a button.

Magnitude of vector a, | a | =

Any vector n that is in the direction of vector a = i j k

Direction cosines l = m = n =

Determination of unit vector â from the vector components of vector a or any vector n that is in the direction vector a.

To determine unit vector â : Enter the vector components of vector a or any vector n that is in the direction of vector a. Press the Unit Vector button.

Vector a or any vector n that is in the direction of a = 3 i -6 j 8 k

Magnitude of vector a, | a | =

Unit vector â = 2.873478856E-1 i -5.746957711E-1 j 7.662610282E-1 k

Vector a = i j k

Vector a | Vector a | Unit Vector | Figure | Formulae | Reset Values | Close

Solution: 1(a) The unit vector that specifies the direction of vector **s** is given by $\hat{\mathbf{s}} = 0.2873\,\mathbf{i} - 0.5747\,\mathbf{j} + 0.7663\,\mathbf{k}$.

1(b) **Worked Solution**

Solve for direction cosines $[\ell, m, n]$ of vector **s**, where

$$\ell = \cos\alpha = \frac{s_x}{|\mathbf{s}|} = \frac{3}{\sqrt{3^2 + (-6)^2 + 8^2}} = \frac{3}{10.44} = 0.2873,$$

$$m = \cos\beta = \frac{s_y}{|\mathbf{s}|} = \frac{-6}{\sqrt{3^2 + (-6)^2 + 8^2}} = \frac{-6}{10.44} = -0.5747 \text{ and}$$

$$n = \cos\gamma = \frac{s_z}{|\mathbf{s}|} = \frac{8}{\sqrt{3^2 + (-6)^2 + 8^2}} = \frac{8}{10.44} = 0.7663.$$

Vector Algebra Tools Solution

Using the Vector Arithmetic Tool: Enter the given components of vector **s** in the Vector **a** user input fields, and enter 0 in the vector **b** user input fields. Then click the Add **a** + **b** button. The answers will be displayed in the output fields as shown.

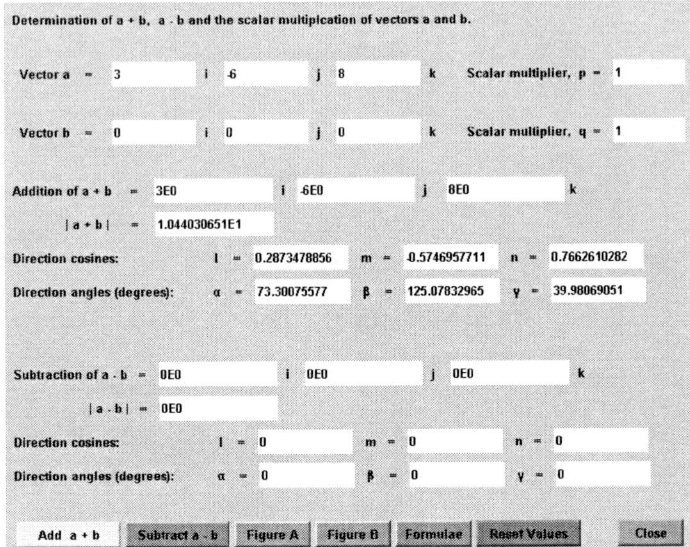

Solution: 1(b) The direction cosines of the displacement vector **s** are given by $\ell = 0.2873$, $m = -0.5747$ and $n = 0.7663$.

1(c) **Worked Solution**

Solve for vector **c**, where

$$\mathbf{c} = \lambda\mathbf{s} = -5(3\,\mathbf{i} - 6\,\mathbf{j} + 8\,\mathbf{k}) = -15\,\mathbf{i} + 30\,\mathbf{j} - 40\,\mathbf{k}.$$

Next solve for the unit vector $\hat{\mathbf{c}}$, where

$$\hat{\mathbf{c}} = \frac{\mathbf{c}}{|\,\mathbf{c}\,|}$$

$$= \frac{-15\,\mathbf{i} + 30\,\mathbf{j} - 40\,\mathbf{k}}{\sqrt{(-15)^2 + 30^2 + (-40)^2}}$$

$$= -0.2873\,\mathbf{i} + 0.5747\,\mathbf{j} - 0.7663\,\mathbf{k}.$$

Vector Algebra Tools Solution

Using the Vector Arithmetic Tool: To determine the displacement vector **c**, enter the components of **s** in the Vector **a** user input fields, together with a value of −5 in the Scalar multiplier p user input field.

Enter 0 in the vector **b** user input fields. Then click the Add **a** + **b** button. The answers will be displayed in the output fields as shown.

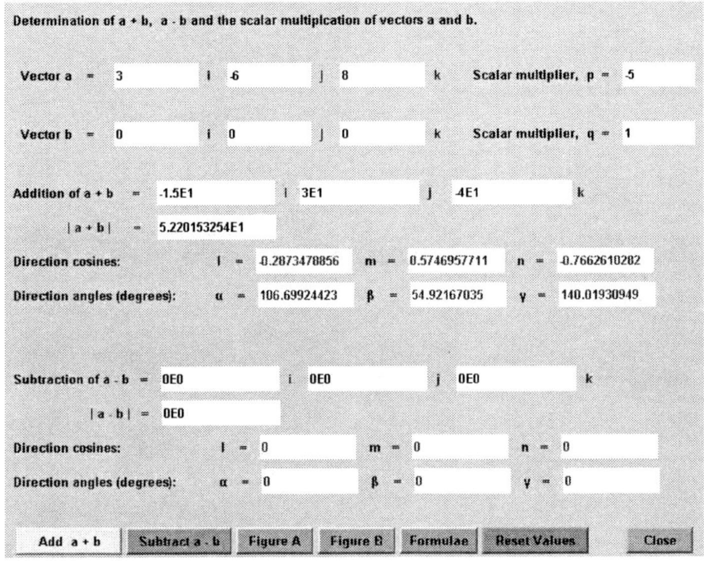

Solution: $\mathbf{c} = -15\,\mathbf{i} + 30\,\mathbf{j} - 40\,\mathbf{k}$ m.

Using the Vector Composition Tool: Enter the calculated components of vector **c** in the Vector **a** or any vector **n** that is in the direction of **a** user input fields. Then click the Unit Vector button. The answers will be displayed in the output fields as shown.

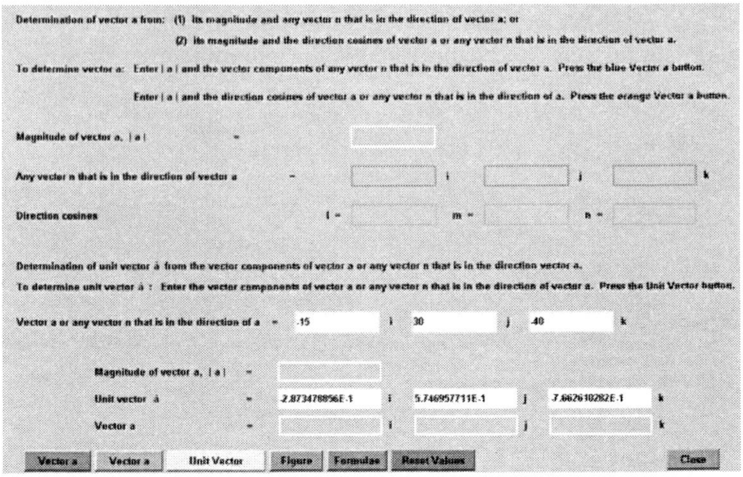

Determination of vector a from: (1) its magnitude and any vector n that is in the direction of vector a; or

(2) its magnitude and the direction cosines of vector a or any vector n that is in the direction of vector a.

To determine vector a: Enter | a | and the vector components of any vector n that is in the direction of vector a. Press the blue Vector a button.

Enter | a | and the direction cosines of vector a or any vector n that is in the direction of a. Press the orange Vector a button.

Magnitude of vector a, | a | =

Any vector n that is in the direction of vector a = i j k

Direction cosines l = m = n =

Determination of unit vector â from the vector components of vector a or any vector n that is in the direction vector a.

To determine unit vector â : Enter the vector components of vector a or any vector n that is in the direction of vector a. Press the Unit Vector button.

Vector a or any vector n that is in the direction of a = -15 i 30 j 40 k

Magnitude of vector a, | a | =

Unit vector â = -2.873478856E-1 i 5.746957711E-1 j 7.662610282E-1 k

Vector a = i j k

Vector a | Vector a | Unit Vector | Figure | Formulae | Reset Values | Close

Solution: 1(c) The unit vector of the collinear vector **c** is $\hat{\mathbf{c}} = -0.2873\,\mathbf{i} + 0.5747\,\mathbf{j} + 0.7663\,\mathbf{k}$.

Answers: 1(a) The unit vector, $\hat{\mathbf{s}} = 0.2873\,\mathbf{i} - 0.5747\,\mathbf{j} + 0.7663\,\mathbf{k}$ and 1(b) $\ell = 0.2873$, $m = -0.5747$, and $n = 0.7663$. Note that the directions cosines $[\ell, m, n]$ of the displacement vector **s** are simply the components of the unit vector $\hat{\mathbf{s}}$. 1(c) The collinear vector **c** points in the opposite direction of vector **s** and therefore $\hat{\mathbf{c}} = -0.2873\,\mathbf{i} + 0.5747\,\mathbf{j} - 0.7663\,\mathbf{k}$.

Question 2. A toy rocket is traveling at a speed of 28 m/s in the direction of the vector $3\,\mathbf{i} - 2\,\mathbf{j} + 6\,\mathbf{k}$. Determine the velocity vector **v** of the rocket along the **i**, **j** and **k** unit vectors.

Worked Solution

Solve for **v**, where

$$\mathbf{v} = |\,\mathbf{v}\,|\,\hat{\mathbf{n}} = |\,\mathbf{v}\,|\,\frac{\mathbf{n}}{|\,\mathbf{n}\,|} = 28\left(\frac{3\,\mathbf{i} - 2\,\mathbf{j} + 6\,\mathbf{k}}{\sqrt{3^2 + (-2)^2 + 6^2}}\right)$$

$$= 12\,\mathbf{i} - 8\,\mathbf{j} + 24\,\mathbf{k} \text{ m/s.}$$

Vector Algebra Tools Solution

Using the Vector Composition Tool: Enter the given speed of the toy rocket in the Magnitude of vector **a**, | **a** | user input field. Enter the given components of the direction vector in the any vector **n** that is in the direction of vector **a** user input fields. Then click the blue Vector **a** button. The answers will be displayed in the output fields as shown.

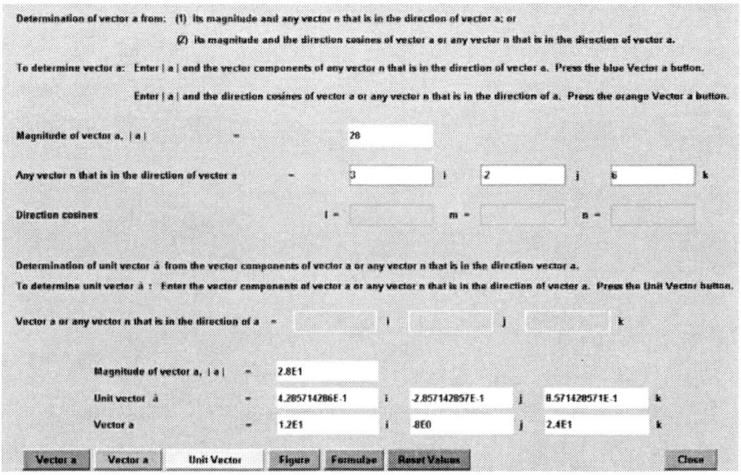

Determination of vector a from: (1) its magnitude and any vector n that is in the direction of vector a; or
(2) its magnitude and the direction cosines of vector a or any vector n that is in the direction of vector a.

To determine vector a: Enter | a | and the vector components of any vector n that is in the direction of vector a. Press the blue Vector a button.
Enter | a | and the direction cosines of vector a or any vector n that is in the direction of a. Press the orange Vector a button.

Magnitude of vector a, | a | = 28

Any vector n that is in the direction of vector a = 3 i -2 j 6 k

Direction cosines l = m = n =

Determination of unit vector â from the vector components of vector a or any vector n that is in the direction vector a.
To determine unit vector â : Enter the vector components of vector a or any vector n that is in the direction of vector a. Press the Unit Vector button.

Vector a or any vector n that is in the direction of a = i j k

Magnitude of vector a, | a | = 2.8E1
Unit vector â = 4.2857142886E-1 i -2.857142857E-1 j 8.571428571E-1 k
Vector a = 1.2E1 i -8E0 j 2.4E1 k

Vector a | Vector a | Unit Vector | Figure | Formulae | Reset Values | Close

Answer: Velocity of the rocket, $\mathbf{v} = 12\,\mathbf{i} - 8\,\mathbf{j} + 24\,\mathbf{k}$ m/s.

Question 3. An object is acted upon by a force **F** that has a magnitude of 13 N and direction cosines given by $\ell = 0.3076923077$, $m = 0.9230769231$ and $n = -0.2307692308$. Determine the force vector **F**.

Worked Solution

Solve for **F**, where

$$\mathbf{F} = |\,\mathbf{F}\,|\,(\ell\,\mathbf{i} + m\,\mathbf{j} + n\,\mathbf{k})$$

$$= 13(0.3076923077\,\mathbf{i} + 0.9230769231\,\mathbf{j} - 0.2307692308\,\mathbf{k})$$

$$= 4\,\mathbf{i} + 12\,\mathbf{j} - 3\,\mathbf{k}\ \text{N}.$$

Vector Algebra Tools Solution

Using the Vector Composition Tool: Enter the given value for the magnitude of vector **F**, | **F** |, in the Magnitude of vector **a**, | **a** | user input field, and enter the given values for the direction cosines [ℓ, m, n] of **F** in the Direction cosines user input fields. Then click the orange Vector **a** button. The answers will be displayed in the output fields as shown.

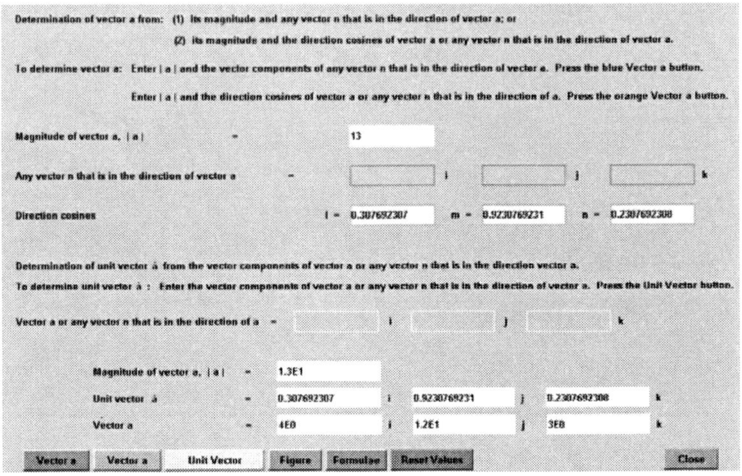

Determination of vector a from: (1) its magnitude and any vector n that is in the direction of vector a; or
(2) its magnitude and the direction cosines of vector a or any vector n that is in the direction of vector a.

To determine vector a: Enter | a | and the vector components of any vector n that is in the direction of vector a. Press the blue Vector a button.
Enter | a | and the direction cosines of vector a or any vector n that is in the direction of a. Press the orange Vector a button.

Magnitude of vector a, | a | = 13

Any vector n that is in the direction of vector a = i j k

Direction cosines l = 0.307692307 m = 0.9230769231 n = 0.2307692308

Determination of unit vector â from the vector components of vector a or any vector n that is in the direction vector a.
To determine unit vector â : Enter the vector components of vector a or any vector n that is in the direction of vector a. Press the Unit Vector button.

Vector a or any vector n that is in the direction of a = i j k

Magnitude of vector a, | a | = 1.3E1
Unit vector â = 0.307692307 i 0.9230769231 j 0.2307692308 k
Vector a = 4E0 i 1.2E1 j 3E0 k

Vector a | Vector a | Unit Vector | Figure | Formulae | Reset Values | Close

Answer: $\mathbf{F} = 4\,\mathbf{i} + 12\,\mathbf{j} - 3\,\mathbf{k}$ N.

Question 4. Determine the direction cosines of the vector that is normal to the plane $\mathbf{r} \bullet (2\,\mathbf{i} - 2\,\mathbf{j} + \mathbf{k}) + 1 = 0$ from the origin.

Worked Solution

Recall that the normal form of the vector equation of a plane is given by

$$\mathbf{r} \bullet \hat{\mathbf{n}} = p,$$

where $\hat{\mathbf{n}}$ is the unit vector normal to the plane from the origin and p is the perpendicular distance of the plane from the origin. Solve for $\hat{\mathbf{n}}$, whose components are the required direction cosines. Since the term on the right hand side of the vector equation must be positive number, we express the vector equation as

$$\mathbf{r} \bullet (-2\,\mathbf{i} + 2\,\mathbf{j} - \mathbf{k}) = 1.$$

Therefore, the vector normal to the plane is $\mathbf{n} = -2\,\mathbf{i} + 2\,\mathbf{j} - \mathbf{k}$ such that the unit vector normal from the origin to the plane is given by

$$\hat{\mathbf{n}} = \frac{\mathbf{n}}{|\mathbf{n}|} = \left(\frac{-2\,\mathbf{i} + 2\,\mathbf{j} - \mathbf{k}}{\sqrt{(-2)^2 + 2^2 + 1^2}} \right) = -\frac{2}{3}\mathbf{i} + \frac{2}{3}\mathbf{j} - \frac{1}{3}\mathbf{k}.$$

Hence, the direction cosines of $\hat{\mathbf{n}}$ are $\left[-\frac{2}{3}, \frac{2}{3}, -\frac{1}{3}\right]$. Note that the vector equation of the plane in normal form is given as

$$\mathbf{r} \bullet \left(-\frac{2}{3}\mathbf{i} + \frac{2}{3}\mathbf{j} - \frac{1}{3}\mathbf{k}\right) = \frac{1}{3}.$$

Vector Algebra Tools Solution

Using the Vector Composition Tool: Enter the components of vector **n** in the Vector **a** or any vector **n** that is in the direction of **a** user input fields. Then click the Unit Vector button. The answers will be displayed in the output fields as shown.

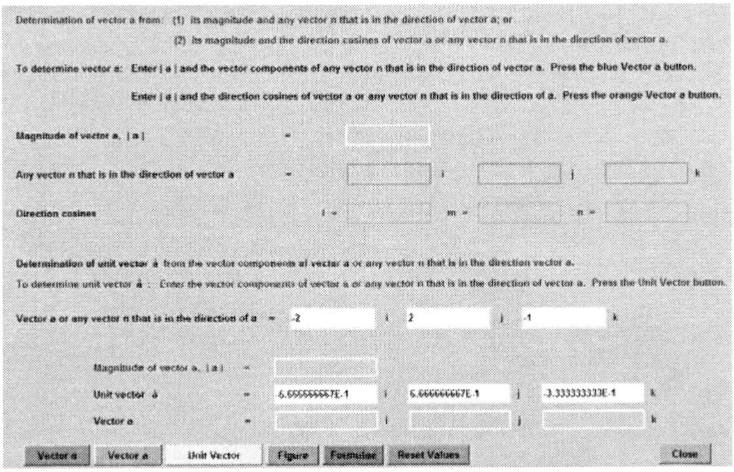

Answer: The direction cosines of the vector normal to the plane from the origin are $\ell = -0.6666$ $(-2/3)$, $m = 0.6666$ $(2/3)$ and $n = -0.3333$ $(-1/3)$.

Question 5. Determine a unit vector that is in the direction perpendicular to both the vectors $\mathbf{a} = 4\,\mathbf{i} + \mathbf{j} + 5\,\mathbf{k}$ and $\mathbf{b} = 3\,\mathbf{i} - 2\,\mathbf{j} + 3\,\mathbf{k}$.

Worked Solution

First determine the vector **n** that is perpendicular to both **a** and **b** given by the vector product $\mathbf{a} \times \mathbf{b}$.

$$\mathbf{n} = \mathbf{a} \times \mathbf{b} = \begin{vmatrix} \mathbf{i} & \mathbf{j} & \mathbf{k} \\ a_x & a_y & a_z \\ b_x & b_y & b_z \end{vmatrix} = \begin{vmatrix} \mathbf{i} & \mathbf{j} & \mathbf{k} \\ 4 & 1 & 5 \\ 3 & -2 & 3 \end{vmatrix} = 13\,\mathbf{i} + 3\,\mathbf{j} - 11\,\mathbf{k}.$$

The corresponding unit vector in the direction of **n** is given by

$$\hat{\mathbf{n}} = \frac{\mathbf{n}}{|\mathbf{n}|} = \frac{13\mathbf{i} + 3\mathbf{j} - 11\mathbf{k}}{\sqrt{13^2 + 3^2 + (-11)^2}} = 0.7518\,\mathbf{i} + 0.1735\,\mathbf{j} - 0.6361\,\mathbf{k}.$$

Note that the unit vector in the opposite direction given by the vector $\mathbf{b} \times \mathbf{a}$ is $\hat{\mathbf{n}} = -0.7518\,\mathbf{i} - 0.1735\,\mathbf{j} + 0.6361\,\mathbf{k}$.

Vector Algebra Tools Solution

Using the Vector Product Tool: Enter the given components of vector **a** in the Vector **a** user input fields and the given components of vector **b** in the Vector **b** user input fields. Then click the Vector Product **a × b** button. The answers will be displayed in the output fields as shown.

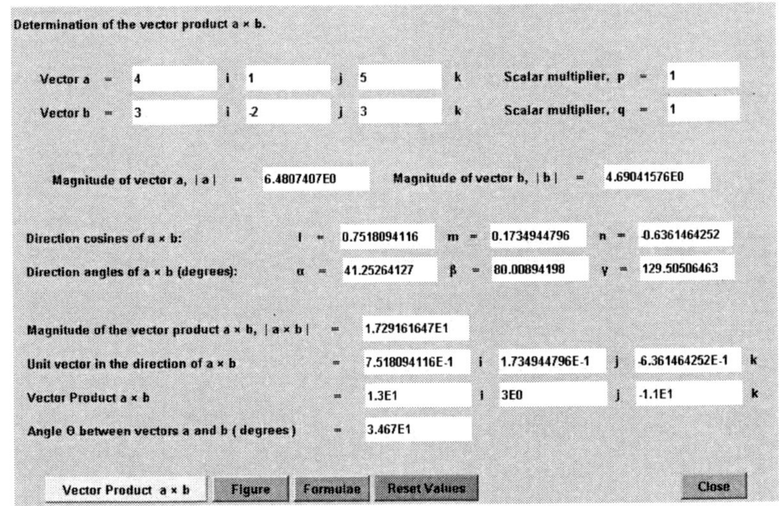

Solution: A unit vector $\hat{\mathbf{n}}$ that is in the direction of the vector $\mathbf{a} \times \mathbf{b}$ is given by $\hat{\mathbf{n}} = 0.7518\,\mathbf{i} + 0.1735\,\mathbf{j} - 0.6361\,\mathbf{k}$.

Note that you can check your answer using the Vector Composition Tool as follows:

Using the Vector Composition Tool: Enter the calculated components of the vector **n** in the Vector **a** or any vector **n** that is in the direction of **a** user input fields. Click the Unit Vector button. The answers will be displayed in the output fields as shown.

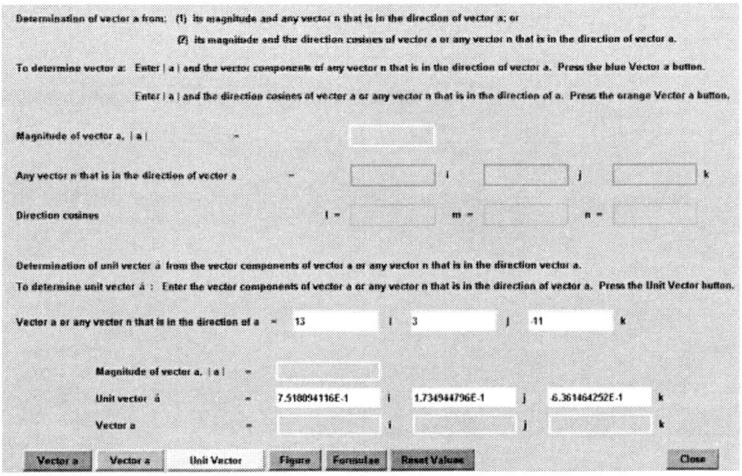

Solution: A unit vector $\hat{\mathbf{n}}$ that is in the direction of vector $\mathbf{a} \times \mathbf{b}$ is given by $\hat{\mathbf{n}} = 0.7518\ \mathbf{i} + 0.1735\ \mathbf{j} - 0.6361\ \mathbf{k}$.

Answer: A unit vector $\hat{\mathbf{n}}$ that is in the direction of the vector $\mathbf{a} \times \mathbf{b}$ is given by $\hat{\mathbf{n}} = 0.7518\ \mathbf{i} + 0.1735\ \mathbf{j} - 0.6361\ \mathbf{k}$. Note that another unit vector perpendicular to the vectors **a** and **b** that is in the direction of $\mathbf{b} \times \mathbf{a}$ is given by $\hat{\mathbf{n}} = -0.7518\ \mathbf{i} - 0.1735\ \mathbf{j} + 0.6361\ \mathbf{k}$.

2.6 NOTES:

CHAPTER 3

VECTOR COMPONENTS IN TWO AND THREE DIMENSIONS

This chapter introduces the properties of a vector that is expressed in terms of its magnitude and direction in a two-dimensional (2D) and three-dimensional (3D) Cartesian coordinate system. Also presented is the method of converting such vectors into their corresponding component form. Physical applications and worked examples in the 2D and 3D Cartesian systems are also included. Furthermore, this chapter demonstrates the use of the 2D and 3D Vector Components software tools in solving the example problems.

3.1 DETERMINATION OF THE COMPONENT FORM OF A VECTOR IN THE TWO-DIMENSIONAL CARTESIAN *X*-*Y* PLANE

In the Cartesian *x*-*y* plane, a **position vector a** that extends from an initial point O, defined at the origin, to the point P, having the Cartesian coordinates (a_x, a_y), can be can be expressed as the sum of the vectors $a_x\,\mathbf{i}$ and $a_y\,\mathbf{j}$ that are directed along the mutually perpendicular positive *x* and *y*-axes, respectively. See figure 3.1. These two vectors are the **Cartesian vector components** of vector **a** and are more commonly referred to as the **vector components** or **vector resolutes** of vector **a**, resulting from the vector projections of vector **a** along the coordinate axes in the directions of the **i** and **j** unit vectors. Also note that vector **a** and the vectors components, $a_x\,\mathbf{i}$ and $a_y\,\mathbf{j}$, are coplanar vectors.

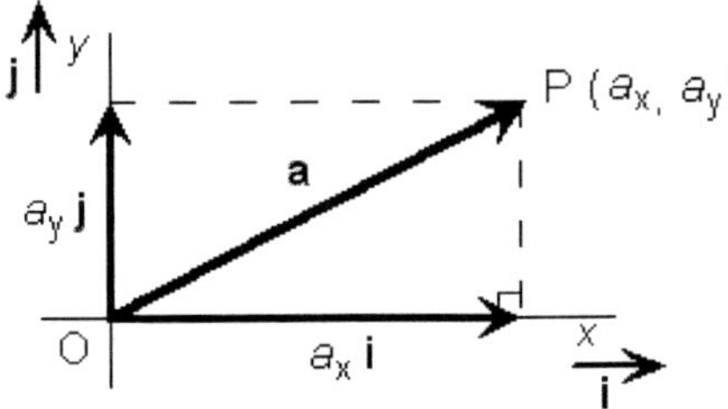

Fig. 3.1. Resolution of vector **a** into its vector components $a_x\,\mathbf{i}$ and $a_y\,\mathbf{j}$ that are directed along the mutually perpendicular positive *x* and *y*-axes, respectively.

Therefore, vector **a** in the Cartesian x-y plane can be expressed as the sum of its vector components $a_x\mathbf{i}$ and $a_y\mathbf{j}$ and can be written in **component form** as

$$\mathbf{a} = a_x\mathbf{i} + a_y\mathbf{j},$$

where a_x and a_y are the **Cartesian rectangular components** of vector **a**, and are more commonly referred to as the **components** or **resolutes** of vector **a** along the positive x and y-axes, respectively. Specifically, a_x and a_y are positive or negative numbers that result from the scalar projections of vector **a** along the coordinate axis in the directions of the **i** and **j** unit vectors, respectively, such that

$$a_x = \mathbf{a} \bullet \mathbf{i} \text{ and } a_y = \mathbf{a} \bullet \mathbf{j}.$$

Therefore, depending on the orientation of vector **a**, the components a_x and a_y can be positive or negative. For example, a_x is positive when it is directed along the positive x-axis and negative when it is directed along the negative x-axis. The corresponding vector components of vector **a** are given by $a_x\mathbf{i}$ and $-a_x\mathbf{i}$, respectively. Also note that the components of the position vector **a** are simply the Cartesian coordinates of the point P in figure 3.1.

Alternatively, vector **a** can also be expressed in terms of its magnitude, $|\mathbf{a}|$, and direction specified by the angle θ, i.e., a positive angle in standard position which by convention is measured in a counterclockwise direction from the positive x-axis in the x-y plane, as shown in figure 3.2. In this type of vector representation, the terminal point P of the position vector **a** is defined by its **plane polar coordinates** $(|\mathbf{a}|, \theta)$ in the Cartesian x-y plane.

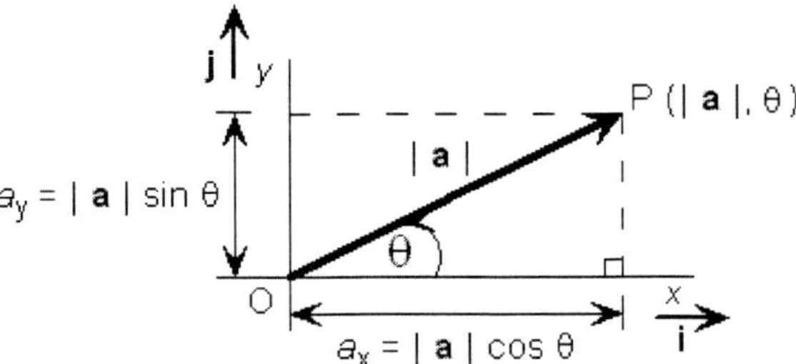

Fig. 3.2. Vector **a** expressed in terms of its magnitude $|\mathbf{a}|$ and the angle θ that is measured in a counterclockwise direction from the positive x-axis.

Referring to figure 3.2 and using elementary trigonometric functions, the components a_x and a_y of vector **a** are related to the magnitude, | **a** |, and the angle θ of vector **a** by the following expressions:

$$a_x = |\mathbf{a}| \cos\theta \text{ and}$$

$$a_y = |\mathbf{a}| \cos(90° - \theta) = |\mathbf{a}| \sin\theta.$$

Therefore, any nonzero two-dimensional vector **a** in the Cartesian *x-y* plane can be expressed in terms of its magnitude and the angle θ in component form as

$$\mathbf{a} = |\mathbf{a}| \cos\theta\, \mathbf{i} + |\mathbf{a}| \sin\theta\, \mathbf{j},$$

where the angle θ determines the direction of vector **a**. Note that in figure 3.2 the components a_x and a_y of vector **a** form two sides of a right angle triangle, whose hypotenuse is the magnitude | **a** | of vector **a**. Hence, the magnitude of vector **a** and the angle θ that vector **a** makes with the positive *x*-axis can be expressed as follows:

$$|\mathbf{a}| = a = \sqrt{a_x^{\,2} + a_y^{\,2}} \text{ and } \theta = \tan^{-1}\left(\frac{a_y}{a_x}\right),$$

where θ is in the range $0° \leq \theta \leq 360°$. The angle θ also can be expressed by the following equivalent equations:

$$\theta = \sin^{-1}\left(\frac{a_y}{|\mathbf{a}|}\right) \text{ and } \theta = \cos^{-1}\left(\frac{a_x}{|\mathbf{a}|}\right).$$

The angle θ is determined by the quadrant of the *x-y* plane in which vector **a** lies and is given by the signs of the components a_x and a_y. See figure 3.3, which summarizes the signs of the components and the corresponding equations for the angle θ for each quadrant of the *x-y* plane in which vector **a** may lie. For example, if vector **a** lies in the second quadrant, where θ = 100°, then a_x is negative, a_y is positive and θ = $180° - \tan^{-1}\left(\left|\frac{a_y}{-a_x}\right|\right)$. Likewise, if vector **a** lies in the fourth quadrant, where θ = 290°, then a_x is positive, a_y is negative and θ = $360° - \tan^{-1}\left(\left|\frac{-a_y}{a_x}\right|\right)$.

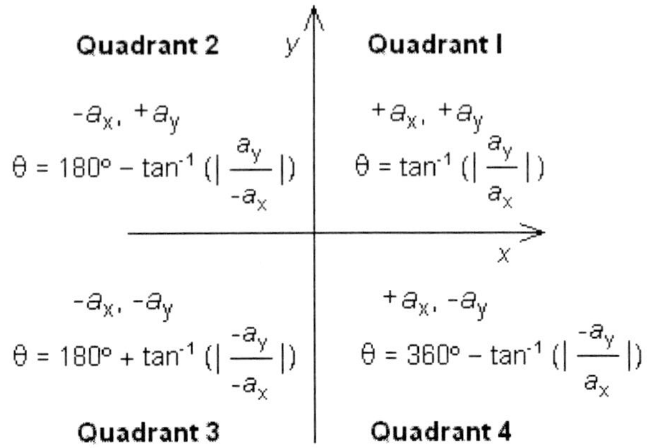

Fig. 3.3. The angle θ is determined by the quadrant of the x-y plane in which vector **a** lies and is given by the signs of the components a_x and a_y.

Therefore, the angle θ can be expressed in terms of the reference angle φ, where $\phi = \tan^{-1}(|\frac{a_y}{a_x}|)$, such that, depending upon the quadrant in which the angle θ lies,

$$\theta = \phi \text{ or } \theta = 180° \pm \phi \text{ or } \theta = 360° - \phi.$$

The **direction cosines** $[\ell, m]$ of vector **a** are given by

$$\ell = \cos\alpha = \frac{a_x}{|\mathbf{a}|} \quad \text{and} \quad m = \cos\beta = \frac{a_y}{|\mathbf{a}|},$$

where the angles α and β are the **direction angles** that vector **a** makes with the positive x and y-axes, respectively, as shown in figure 3.4. The direction angles α and β are, by convention, in the range from 0° to 180° and are given by

$$\alpha = \cos^{-1}(\frac{a_x}{|\mathbf{a}|}) \text{ and } \beta = \cos^{-1}(\frac{a_y}{|\mathbf{a}|}).$$

Depending on the signs of the components of vector **a**, the corresponding direction angles α and β are acute when the components of vector **a** are positive and are obtuse when the components of vector **a** are negative.

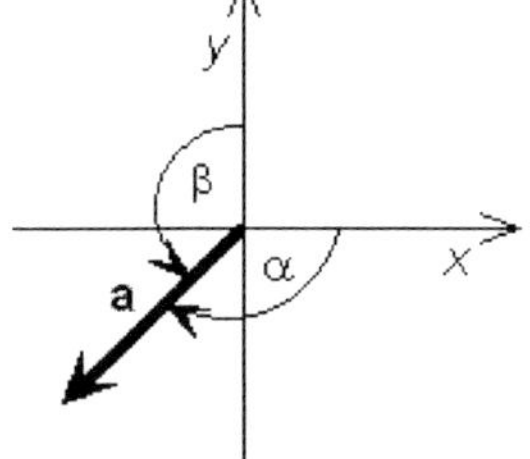

Fig. 3.4. Direction angles α and β that vector **a** makes with the positive *x* and *y*-axes, respectively, in a two-dimensional Cartesian plane.

3.2 DETERMINATION OF THE COMPONENT FORM OF A VECTOR IN THE THREE-DIMENSIONAL CARTESIAN *X-Y-Z* PLANE

In a three-dimensional Cartesian coordinate system, if **i**, **j** and **k** are the unit vectors directed along the positive *x*, *y* and *z*-axes, respectively, then the **position vector a** drawn from the origin, O, to the point P, having the Cartesian coordinates (a_x, a_y, a_z), as shown in figure 3.5, can be expressed in component form as the sum of the **vector components** a_x **i**, a_y **j**, and a_z **k** that are directed parallel to the mutually perpendicular positive *x*, *y* and *z*-axes, respectively. Thus, position vector **a** can be written in component form as

$$\mathbf{a} = a_x\,\mathbf{i} + a_y\,\mathbf{j} + a_z\,\mathbf{k}.$$

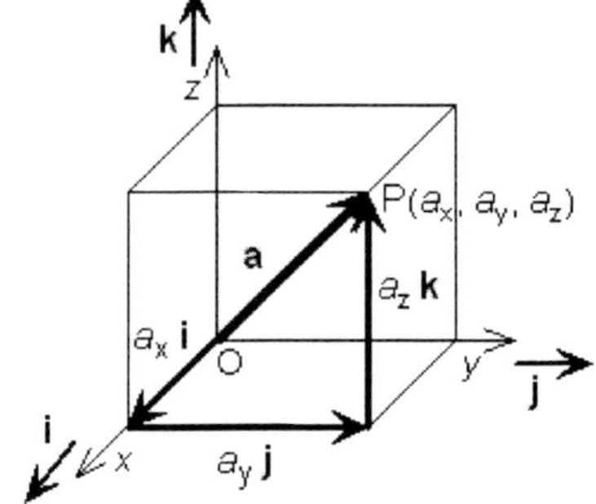

Fig. 3.5. The vector components of the position vector **a** are the vectors a_x **i**, a_y **j** and a_z **k** that are parallel to the *x*, *y* and *z*-axes, respectively.

The **components** or **resolutes** of vector **a** are the scalar projections of vector **a** along each of the three coordinate axes that are in the directions of the **i**, **j** and **k** unit vectors, and can be expressed by the following equations:

$$a_x = |\mathbf{a}| \cos\alpha, \quad a_y = |\mathbf{a}| \cos\beta \quad \text{and} \quad a_z = |\mathbf{a}| \cos\gamma,$$

where the angles α, β and γ are the **direction angles** that vector **a** makes with the positive *x*, *y* and *z*-axes, respectively. These components have been determined by applying elementary trigonometry principles to a right angle triangle that has been constructed using the magnitude of **a** and each of the corresponding directions angles α, β and γ, as shown in figure 3.6.

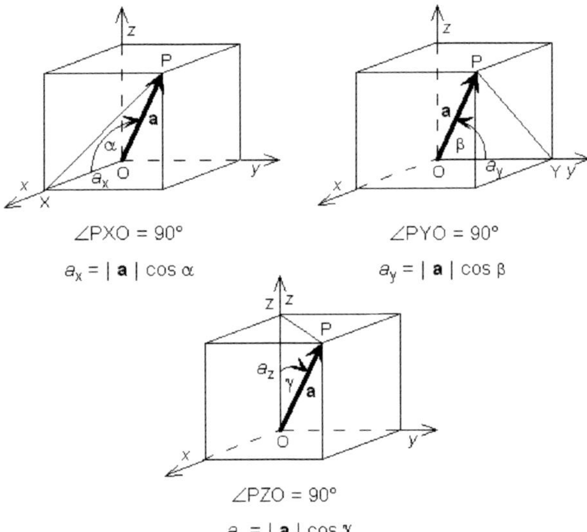

Fig. 3.6. Components of vector **a** expressed in terms their magnitude, $|\mathbf{a}|$, and the cosines of each of the direction angles α, β and γ.

The cosines of the direction angles that vector **a** makes with each coordinate axis are the **direction cosines** denoted by [ℓ, *m*, *n*], and can be expressed as

$$\ell = \cos\alpha = \frac{a_x}{|\mathbf{a}|}, \quad m = \cos\beta = \frac{a_y}{|\mathbf{a}|} \quad \text{and} \quad n = \cos\gamma = \frac{a_z}{|\mathbf{a}|},$$

where the magnitude of vector **a** is given by

$$|\mathbf{a}| = \sqrt{a_x^{\,2} + a_y^{\,2} + a_z^{\,2}}.$$

Therefore, any nonzero three-dimensional vector **a** can be expressed as the product of its magnitude and a unit vector that is in the direction of vector **a**, such that,

$$\mathbf{a} = |\,\mathbf{a}\,|\,\hat{\mathbf{a}} = |\,\mathbf{a}\,|\,(\ell\,\mathbf{i} + m\,\mathbf{j} + n\,\mathbf{k}) = |\,\mathbf{a}\,|\,(\cos\alpha\,\mathbf{i} + \cos\beta\,\mathbf{j} + \cos\gamma\,\mathbf{k}),$$

where the components of unit vector $\hat{\mathbf{a}}$ are the directions cosines of vector **a**.

Since the direction cosines are not independent, they must satisfy the relational condition

$$\cos^2\alpha + \cos^2\beta + \cos^2\gamma = 1.$$

Thus, using the above identity you can determine any one of the three direction angles α, β or γ for vector **a** if the other two direction angles are known. For example, if the direction angles α and β are known, then the direction angle γ that vector **a** makes with respect to the positive z-axis can be determined using the following expression:

$$\gamma = \cos^{-1}\left(\pm\sqrt{1 - \cos^2\alpha - \cos^2\beta}\right).$$

Note that a positive value for the argument of the inverse cosine function corresponds to a component of vector **a** along the positive z-axis, and a negative value for the argument corresponds to a component along the negative z-axis. Likewise, given the direction angles α and γ, you can determine the direction angle β; and given the direction angles β and γ, you can determine the direction angle α. Once all three of the directions angles are known, you can determine each of the corresponding direction cosines of vector **a**, from which you can then construct its unit vector. Thus, vector **a** can be expressed in component form as the product its magnitude and unit vector, whose components are the direction cosines of vector **a**. For example, if the displacement vector **s** has a magnitude $|\,\mathbf{s}\,| = 20$ m and two of its three direction angles are given as $\alpha = 60°$ and $\beta = 45°$, then, after determining the third direction angle γ ($\gamma = 60°$ or $120°$), vector **a** can be expressed as $20(\cos 60°\,\mathbf{i} + \cos 45°\,\mathbf{j} \pm \cos 60°\,\mathbf{k})$ m. The corresponding x, y and z components of vector **s** are $s_x = 20\cos 60°$ m, $s_y = 20\cos 45°$ m and $s_z = \pm 20\cos 60°$ m, respectively. Note that there are two possible values of γ and therefore two possible directions that define the line that is parallel to **s**. In order to know which of the two possible angles of γ that correctly indicates the direction of the vector **s** additional information is required. Typically, this information is given in the problem that you are trying to solve.

For examples, see Questions 1 and 2 using the 3D Vector Components Tool in the Worked Examples section of this Chapter.

3.3 PROBLEM SOLVING APPLICATIONS USING COMPONENTS AND VECTOR COMPONENTS OF A VECTOR

In a Cartesian coordinate system, a vector can be expressed in terms of its magnitude and its direction defined by the angle(s) that the vector makes with the reference coordinate axis (or axes). Although, this angular representation is used to depict a vector geometrically, it is more convenient to convert a vector expressed in this manner into its component form to solve vector algebra problems. The following examples demonstrate the various applications of this conversion in solving both two and three-dimensional vector algebra problems.

- Determination of the x and y components of a displacement vector **s**, where **s** is expressed in terms of its distance and direction defined by the angle θ between the positive x-axis and **s** in the x-y plane.

- Determination of the x and y components of a velocity vector **v**, where **v** is expressed in terms of its speed and direction defined by the angle θ between the positive x-axis and **v** in the x-y plane.

- Determination of the component form of any nonzero three-dimensional vector, where the vector is expressed in terms of its magnitude and *any* two of its three directions angles (or *any* two of its direction cosines) in the x-y-z plane.

- Determination of the x and y components of the velocity and displacement vectors of a projectile in the x-y plane. For example, as shown in figure 3.7, the velocity of a projectile, **v**, which is tangential to the path of the projectile, can be expressed in terms of its horizontal velocity component, v_x, and its vertical velocity component, v_y, at any time t, by the following equations:

$$v_x = v_{xo} = \mathbf{v}_o \bullet \mathbf{i} = |\,\mathbf{v}_o\,| \cos\theta = \text{a constant} \quad \text{and}$$

$$v_y = v_{yo} - gt = |\,\mathbf{v}_o\,| \sin\theta - gt,$$

such that

$$\mathbf{v} = v_x\,\mathbf{i} + v_y\,\mathbf{j} = |\,\mathbf{v}_o\,| \cos\theta\,\mathbf{i} + (|\,\mathbf{v}_o\,| \sin\theta - gt)\,\mathbf{j},$$

where $\mathbf{v}_o$ is the initial velocity of the projectile and g is the magnitude of the acceleration due to gravity.

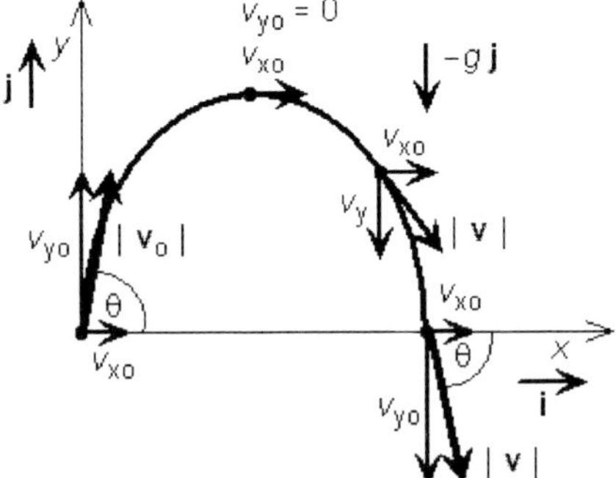

Fig. 3.7. A projectile launched from the origin at an angle θ above the horizontal with a speed of $|\mathbf{v}_o|$.

- Determination of the x and y components of the forces acting on a mass involving Newton's second law of motion, given by $\mathbf{F} = m\mathbf{a}$, in the Cartesian x-y plane. For example, as shown in figure 3.8, the single force $\mathbf{F}$ acting on an object of mass m that moves with an acceleration $\mathbf{a}$ in the direction of the positive x-axis can be expressed in terms of the x component of $\mathbf{F}$, F_x, such that

$$F_x = \mathbf{F} \bullet \mathbf{i} = |\mathbf{F}| \cos\theta = ma_x,$$

where a_x is the x component of $\mathbf{a}$ along the x-axis.

The y component of $\mathbf{F}$, F_y, is given by

$$F_y = |\mathbf{F}| \sin\theta + |\mathbf{N}| - |\mathbf{W}| = 0,$$

where $\mathbf{N}$ is the normal force perpendicular to the surface and $\mathbf{W}$ is the weight of the object, i.e., the gravitational force on the object, and is given by

$$\mathbf{W} = -mg\,\mathbf{j},$$

where g is the acceleration due to gravity of the object in free-fall.

Thus, since $F_y = 0$,

$$\mathbf{F} = F_x\mathbf{i} = (|\mathbf{F}| \cos\theta)\,\mathbf{i} = ma_x\mathbf{i}.$$

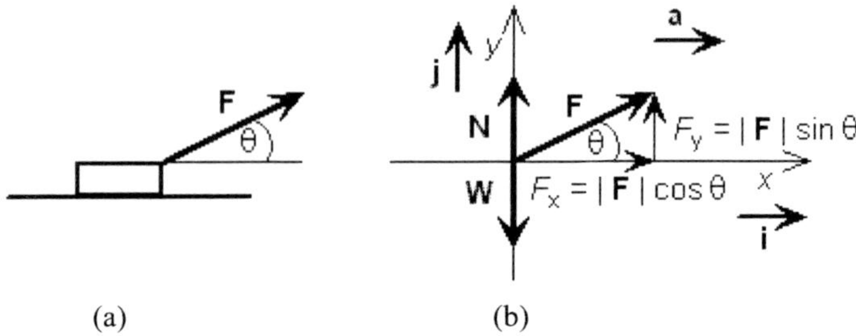

Fig. 3.8. (a) Force vector **F** inclined at an angle θ to the horizontal acting on a mass *m* that accelerates along the horizontal plane. (b) Free-body diagram that represents the external forces on the mass.

3.4 WORKED EXAMPLES

3.4.1 Components and vector components in the 2D Cartesian *x*-*y* plane

The **2D Vector Components Tool** can be used to determine the component form of a vector given its magnitude and its direction defined by the angle θ that is measured in a counterclockwise direction from the positive *x*-axis in the Cartesian *x*-*y* plane. Also calculated are the vector's direction angles, α and β, and the unit vector that is in the direction of the vector. See Appendix B for information on how to enter values into any of the Vector Algebra Tools.

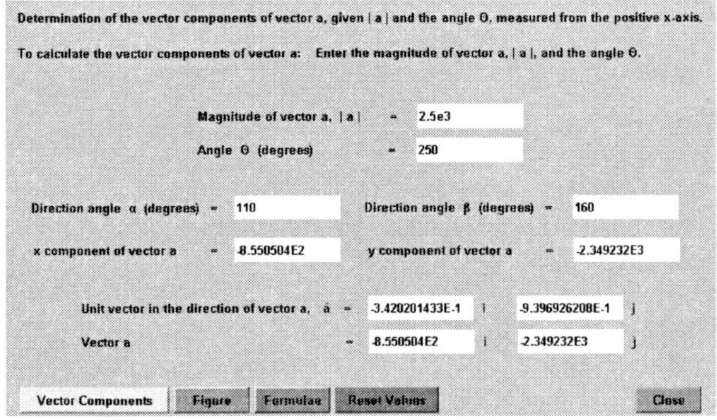

Fig. 3.9. The 2D Vector Components Tool from the Vector Algebra Tools Software Program.

Question 1. Given the magnitude and the direction of the force vectors **F**, **F′** and **F″**, as shown in figure 3.10, determine the corresponding x and y components of each force. The unit of measurement of each force is the newton.

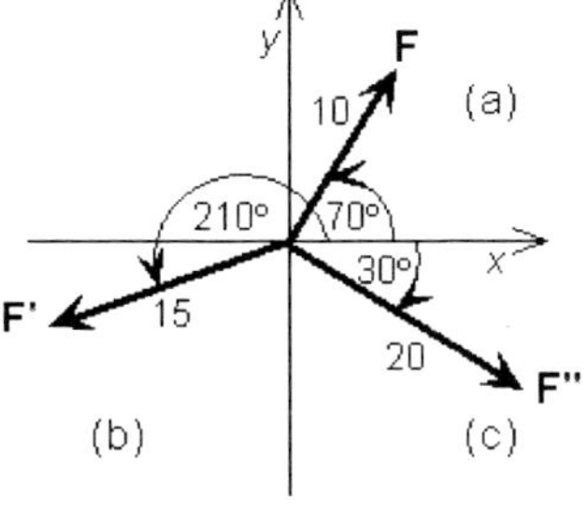

Fig. 3.10.

Worked Solutions

Vector **F**: In quadrant (a) solve for the x and y components of the force vector **F** along the x and y-axes, respectively, where

$$F_x = |\,\mathbf{F}\,| \cos\theta = 10 \cos 70° = 3.4202 \text{ N} \text{ and}$$

$$F_y = |\,\mathbf{F}\,| \sin\theta = 10 \sin 70° = 9.3969 \text{ N}.$$

Vector Algebra Tools Solutions

Using the 2D Vector Components Tool: Enter the given magnitude for vector **F**, | **F** |, in the Magnitude of vector **a**, | **a** | user input field. Next enter the given angle vector **F** makes with respect to the positive x-axis in the Angle θ user input field. Then click the Vector Components button. The answers will be displayed in the output fields as shown.

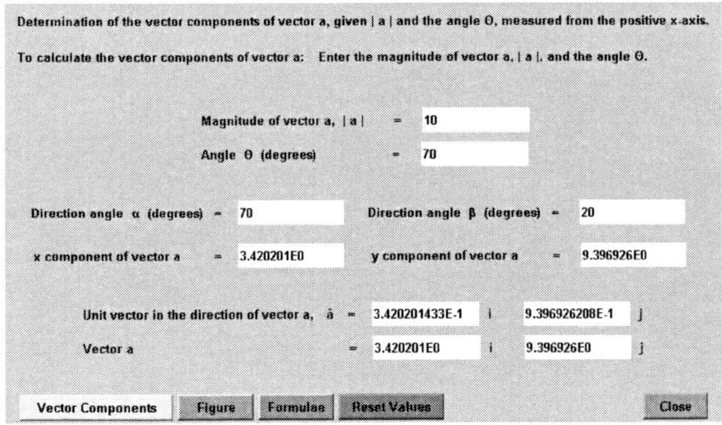

Solution for Quadrant (a): $F_x = 3.4202$ N and $F_y = 9.3969$ N.

Worked Solutions

Vector **F**′: In quadrant (b) solve for the x and y components of the force vector **F**′ along the x and y-axes, respectively, where

$$F'_x = |\,\mathbf{F}'\,| \cos\theta = 15 \cos 210° = -12.9904 \text{ N} \quad \text{and}$$

$$F'_y = |\,\mathbf{F}'\,| \sin\theta = 15 \sin 210° = -7.5 \text{ N}.$$

Vector Algebra Tools Solution

Using the 2D Vector Components Tool: Enter the given magnitude for vector **F**′, | **F**′ |, in the Magnitude of vector **a**, | **a** | user input field. Next enter the given angle that vector **F**′ makes with respect to the positive x-axis in the Angle θ user input field. Then click the Vector Components button. The answers will be displayed in the output fields as shown.

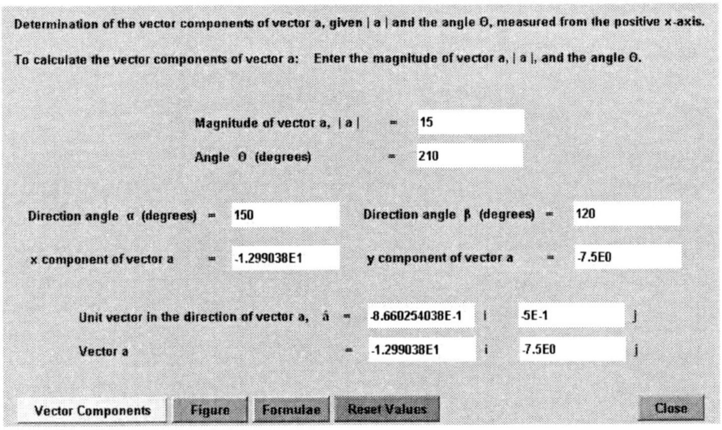

Determination of the vector components of vector a, given | a | and the angle Θ, measured from the positive x-axis.

To calculate the vector components of vector a: Enter the magnitude of vector a, | a |, and the angle Θ.

Magnitude of vector a, | a | = 15

Angle Θ (degrees) = 210

Direction angle α (degrees) = 150 Direction angle β (degrees) = 120

x component of vector a = -1.299038E1 y component of vector a = -7.5E0

Unit vector in the direction of vector a, â = -8.660254038E-1 i -5E-1 j

Vector a = -1.299038E1 i -7.5E0 j

Vector Components Figure Formulae Reset Values Close

Solution for Quadrant (b): $F'_x = -12.9904$ N and $F'_y = -7.5$ N.

Worked Solutions

Vector **F**″: In quadrant (c) solve for the x and y components of the force vector **F**″ along the x and y-axes, respectively, where

$$F''_x = |\,\mathbf{F}''\,| \cos\theta = 20 \cos 330° = 17.3205 \text{ N} \quad \text{and}$$

$$F''_y = |\,\mathbf{F}''\,| \sin\theta = 20 \sin 330° = -10.0 \text{ N}.$$

Vector Algebra Tools Solution

Using the 2D Vector Components Tool: Enter the given magnitude for vector **F**″, | **F**″ |, in the Magnitude of vector **a**, | **a** | user input field. Next enter the angle that vector **F″** makes with respect to the positive x-axis in the Angle θ user input field. Then click the Vector Components button. The answers will be displayed in the output fields as shown.

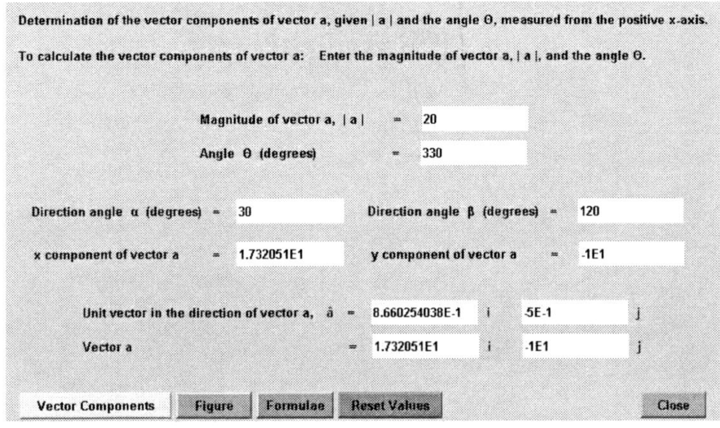

Solution for Quadrant (c): $F''_x = 17.3205$ N and $F''_y = -10$ N.

Answers: 1(a) Components of **F** are $F_x = 3.4202$ N and $F_y = 9.3969$ N, 1(b) Components of **F** ′ are $F'_x = -12.9904$ N and $F'_y = -7.5$ N and 1(c) Components of **F**″ are $F''_x = 17.3205$ N and $F''_y = -10$ N.

Question 2. In figure 3.11, two moored ships are located at the points S_1 and S_2 with corresponding position vectors $\mathbf{s}_1$ and $\mathbf{s}_2$, as measured by an observer at the origin, O. Calculate the distance between the two ships.

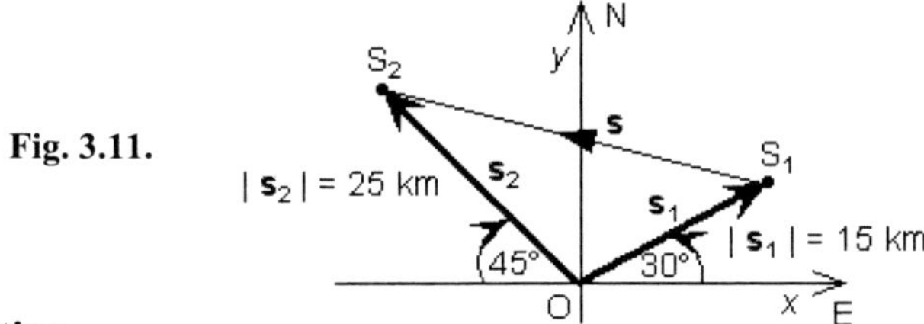

Fig. 3.11.

Worked Solution

The distance between the two ships is given by the magnitude of the displacement vector **s**, where $\mathbf{s} = \mathbf{s}_2 - \mathbf{s}_1$ is the position vector of S_2 relative to S_1, as shown in figure 3.11. In the figure, the positive x-axis points to the east and the positive y-axis points north, and by convention, the direction of each position vector is defined by the angle θ that is measured in a counterclockwise direction from the positive x-axis.

Express $\mathbf{s}_2$ in component form by first solving for the x and y components of $\mathbf{s}_2$ along the x and y-axes, respectively, where

$$s_{2x} = |\,\mathbf{s}_2\,| \cos\theta = 25 \cos 135° = -17.6777 \text{ km} \quad \text{and}$$

$$s_{2y} = |\,\mathbf{s}_2\,| \sin\theta = 25 \sin 135° = 17.6777 \text{ km}.$$

Therefore, $\mathbf{s}_2$ can be written as the sum of its x and y components in the direction of the **i** and **j** unit vectors as

$$\mathbf{s}_2 = s_{2x}\,\mathbf{i} + s_{2y}\,\mathbf{j} = -17.6777\,\mathbf{i} + 17.6777\,\mathbf{j} \text{ km}.$$

Express $\mathbf{s}_1$ in component form by first solving for the x and y components of $\mathbf{s}_1$ along the x and y-axes, respectively, where

$$s_{1x} = |\,\mathbf{s}_1\,| \cos\theta = 15 \cos 30° = 12.9904 \text{ km} \quad \text{and}$$

$$s_{1y} = |\,\mathbf{s}_1\,| \sin\theta = 15 \sin 30° = 7.5 \text{ km}.$$

Therefore, $\mathbf{s}_1$ can be written as the sum of its x and y components in the direction of the **i** and **j** unit vectors as

$$\mathbf{s}_1 = s_{1x}\,\mathbf{i} + s_{1y}\,\mathbf{j} = 12.9904\,\mathbf{i} + 7.5\,\mathbf{j} \text{ km}.$$

Determine the displacement vector **s**, where

$$\mathbf{s} = \mathbf{s}_2 - \mathbf{s}_1 = -17.6777\,\mathbf{i} + 17.6777\,\mathbf{j} - (12.9904\,\mathbf{i} + 7.5\,\mathbf{j})$$

$$= -30.6681\,\mathbf{i} + 10.1777\,\mathbf{j} \text{ km}.$$

Solve for the distance, $|\,\mathbf{s}\,|$, where

$$|\,\mathbf{s}\,| = \sqrt{(-30.6681)^2 + 10.1777^2}$$

$$= 32.31 \text{ km}$$

Note that the distance between the two ships can also be determined by calculating the magnitude of the position vector of S_1 relative to S_2, that is $|\,\mathbf{s}\,| = |\,\mathbf{s}_1 - \mathbf{s}_2\,| = |-(\mathbf{s}_2 - \mathbf{s}_1)\,|$.

Vector Algebra Tools Solution

Using the 2D Vector Components Tool: Enter the given magnitude for vector $\mathbf{s}_2$, | $\mathbf{s}_2$ |, in the Magnitude of vector **a**, | **a** | user input field. Next enter the given angle vector $\mathbf{s}_2$ makes with respect to the positive x-axis in the Angle θ user input field. Then click the Vector Components button. The answers will be displayed in the output fields as shown. Click the Reset Values button before continuing to the next step.

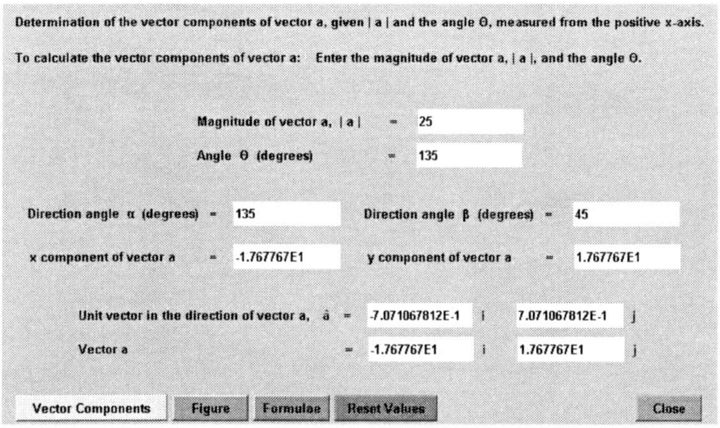

Determination of the vector components of vector a, given | a | and the angle Θ, measured from the positive x-axis.

To calculate the vector components of vector a: Enter the magnitude of vector a, | a |, and the angle Θ.

Magnitude of vector a, | a | = 25

Angle Θ (degrees) = 135

Direction angle α (degrees) = 135 Direction angle β (degrees) = 45

x component of vector a = -1.767767E1 y component of vector a = 1.767767E1

Unit vector in the direction of vector a, â = -7.071067812E-1 i 7.071067812E-1 j

Vector a = -1.767767E1 i 1.767767E1 j

Vector Components | Figure | Formulae | Reset Values | Close

Solution: $\mathbf{s}_2 = -17.6777\,\mathbf{i} + 17.6777\,\mathbf{j}$ km.

Using the 2D Vector Components Tool: Enter the given magnitude for vector $\mathbf{s}_1$, | $\mathbf{s}_1$ |, in the Magnitude of vector **a**, | **a** | user input field. Next enter the given angle vector $\mathbf{s}_1$ makes with respect to the positive x-axis in the Angle θ user input field. Then click the Vector Components button. The answers will be displayed in the output fields as shown.

Determination of the vector components of vector a, given | a | and the angle Θ, measured from the positive x-axis.

To calculate the vector components of vector a: Enter the magnitude of vector a, | a |, and the angle Θ.

Magnitude of vector a, | a | = 15

Angle Θ (degrees) = 30

Direction angle α (degrees) = 30 Direction angle β (degrees) = 60

x component of vector a = 1.299038E1 y component of vector a = 7.5E0

Unit vector in the direction of vector a, â = 8.660254038E-1 i 5E-1 j

Vector a = 1.299038E1 i 7.5E0 j

Vector Components | Figure | Formulae | Reset Values | Close

Solution: $\mathbf{s}_1 = 12.9904\ \mathbf{i} + 7.5\ \mathbf{j}$ km.

Using the Vector Arithmetic Tool: Enter the given components of vector $\mathbf{s}_2$ in the Vector **a** user input fields and the given components of vector $\mathbf{s}_1$ in the Vector **b** user input fields. Then click the Subtract **a** – **b** button. The answers will be displayed in the output fields as shown.

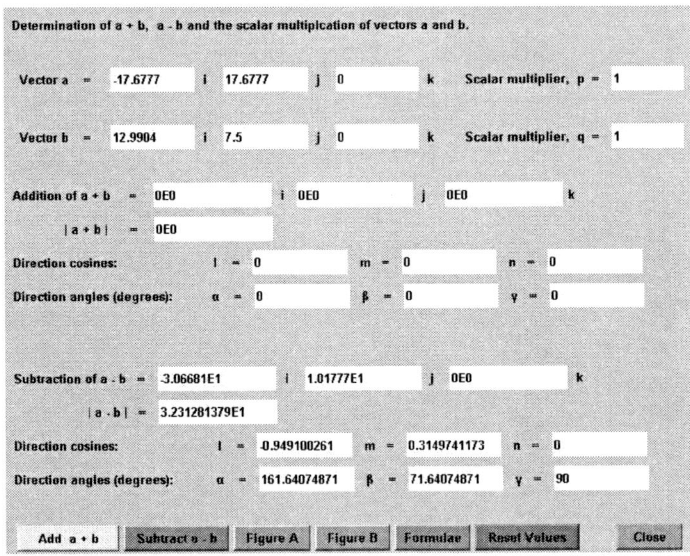

Solutions: $\mathbf{s} = -30.6681\ \mathbf{i} + 10.1777\ \mathbf{j}$ km and $|\ \mathbf{s}\ | = 32.31$ km.

Answer: The distance between the two ships is $|\ \mathbf{s}\ | = 32.31$ km.

Question 3. In figure 3.12 a positive electric charge $q = 4\ \mu C$ (i.e., 4×10^{-6} C) is moving in the *x*-*y* plane with a speed of 2.5×10^4 m/s in a direction that is at an angle of 40° with respect to a uniform external magnetic field given by $\mathbf{B} = 0.55\ \mathbf{i}$ T. Calculate the magnitude of the magnetic force on *q*. *Hint*: The force on the moving electric charge *q* in a uniform magnetic field is given by $\mathbf{F}_B = q(\mathbf{v} \times \mathbf{B})$.

Fig. 3.12.

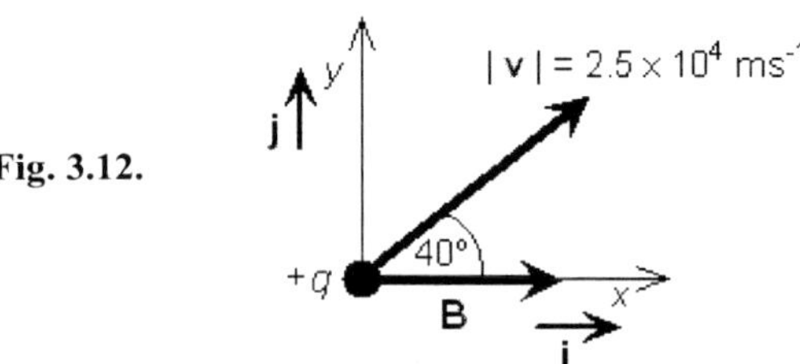

Worked Solution

First solve for the *x* and *y* components of the velocity vector **v** of *q* along the *x* and *y*-axes, respectively, where

$$v_x = |\,\mathbf{v}\,| \cos\theta = 2.5 \times 10^4 \cos 40° = 1.91511 \times 10^4 \text{ m/s and}$$

$$v_y = |\,\mathbf{v}\,| \sin\theta = 2.5 \times 10^4 \sin 40° = 1.60697 \times 10^4 \text{ m/s.}$$

Express **v** in component form as the sum of its components v_x and v_y that are in the direction of the **i** and **j** unit vectors, respectively, where

$$\mathbf{v} = v_x\,\mathbf{i} + v_y\,\mathbf{j} = 19151.1\,\mathbf{i} + 16069.7\,\mathbf{j} \text{ m/s.}$$

Solve for $\mathbf{v} \times \mathbf{B}$, where

$$\mathbf{v} \times \mathbf{B} = \begin{vmatrix} \mathbf{i} & \mathbf{j} & \mathbf{k} \\ v_x & v_y & v_z \\ B_x & B_y & B_z \end{vmatrix}$$

$$= \begin{vmatrix} \mathbf{i} & \mathbf{j} & \mathbf{k} \\ 19151.1 & 16069.7 & 0 \\ 0.55 & 0 & 0 \end{vmatrix}$$

$$= -8.8383 \times 10^3\,\mathbf{k}\ \text{(m/s)T.}$$

The magnetic force, $\mathbf{F}_B$, on the moving charge q, is given by

$$\mathbf{F}_B = q(\mathbf{v} \times \mathbf{B}) = 4 \times 10^{-6}\,(-8.8383 \times 10^3)\,\mathbf{k} = -3.5353 \times 10^{-2}\,\mathbf{k}\ \mathrm{N}.$$

The magnitude of $\mathbf{F}_B$ is $|\,\mathbf{F}_B\,| = 3.5353 \times 10^{-2}$ N.

Vector Algebra Tools Solution

Using the 2D Vector Components Tool: Enter the given speed of q, $|\,\mathbf{v}\,|$, in the Magnitude of vector **a**, $|\,\mathbf{a}\,|$ user input field. Next enter the given angle of q, measured from the positive x-axis, in the Angle θ user input field. Then click the Vector Components button. The answers will be displayed in the output fields as shown.

Determination of the vector components of vector a, given | a | and the angle θ, measured from the positive x-axis.

To calculate the vector components of vector a: Enter the magnitude of vector a, | a |, and the angle θ.

Magnitude of vector a, | a | = 2.5e4

Angle θ (degrees) = 40

Direction angle α (degrees) = 40 Direction angle β (degrees) = 50

x component of vector a = 1.915111E4 y component of vector a = 1.606969E4

Unit vector in the direction of vector a, â = 7.660444431E-1 i 6.427876097E-1 j

Vector a = 1.915111E4 i 1.606969E4 j

Vector Components Figure Formulae Reset Values Close

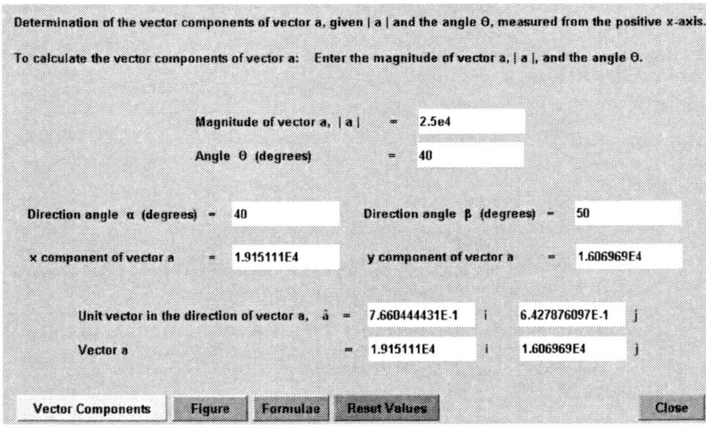

Solution: $\mathbf{v} = 1.91511 \times 10^4\,\mathbf{i} + 1.60697 \times 10^4\,\mathbf{j}$ m/s.

Using the Vector Product Tool: Enter the calculated components of vector **v** in the Vector **a** user input fields. Enter the given components of the uniform external magnetic field **B** in the Vector **b** user input fields. Enter a value of 4×10^{-6} for the positive electric charge q in the Scalar multiplier, p user input field. Then click the Vector Product $\mathbf{a} \times \mathbf{b}$ button. The answers will be displayed in the output fields as shown.

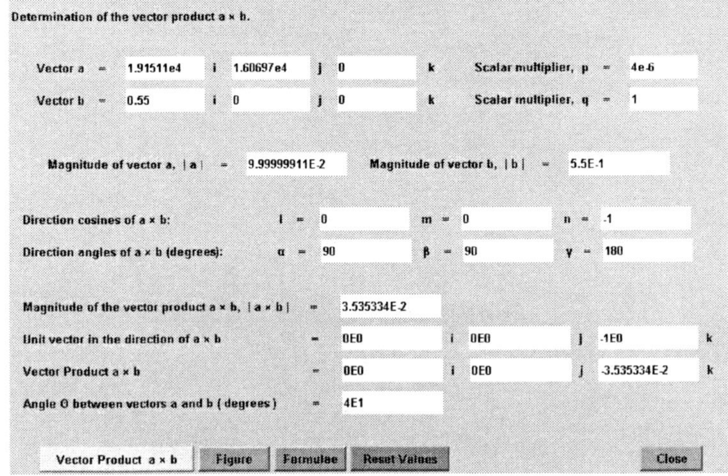

Solutions: $\mathbf{F}_B = -3.5353 \times 10^{-2}\ \mathbf{k}$ N and $|\ \mathbf{F}_B\ | = 3.5353 \times 10^{-2}$ N.

Answer: The magnitude of the magnetic force on the positive charge q is $|\ \mathbf{F}_B\ | = 3.5353 \times 10^{-2}$ N.

3.4.2 Components and vector components in the 3D Cartesian *x*-*y*-*z* plane

The **3D Vector Components Tool** can be used to determine a three-dimensional vector in component form given its magnitude and *any* two of the three direction angles α, β or γ that the vector makes the positive *x*, *y* and *z*-axes of a three-dimensional Cartesian coordinate system, respectively. Also calculated are the third direction angle and the unit vector that is in the direction of the vector whose components are the direction cosines of the vector. See Appendix B for information on how to enter values into any of the Vector Algebra Tools.

Determination of the vector components of vector a given its magnitude | a | and any two of the direction angles, α, β and γ.

To calculate vector a and the third direction angle: Enter | a | and the two known direction angles.

Magnitude of vector a, | a |	=	12
Angle α (degrees)	=	35
Angle β (degrees)	=	75 or 105
Angle γ (degrees)	=	59.7178686

Unit vector in the direction of vector a, â	=	8.1915204433E-1	i	±2.5881904510E-1	j	5.1186192498E-1	k
Vector a	=	9.829825E0	i	±3.105829E0	j	6.142343E0	k

Vector a & Angle γ | Vector a & Angle β | Vector a & Angle α | Figure | Formulae | Reset Values | Close

Fig. 3.13. The 3D Vector Components Tool from the Vector Algebra Tools Software Program.

Question 1. The electric field vector **E** acts on a positive charge of 6.75 μC located at the origin of a three-dimensional Cartesian coordinate system, as shown in figure 3.14. If **E** is inclined with respect to the positive *x*, *y* and *z*-axes as shown, solve for (a) the angle γ that the electric field vector **E** makes with the positive *z*-axis, (b) the electric field vector **E** and (c) the magnitude of the force **F** acting on the electric charge, given that α = 120°, β = 45° and | **E** | = 200 N/C. Assume that the component of the electric field vector **E** along the *z*-axis is positive.

Worked Solutions

1(a) We have $\ell = \cos\alpha = \cos 120° = -0.5$, $m = \cos\beta = \cos 45° = 0.7071$ and $n = \cos\gamma$. Solve for the direction angle γ that **E** makes with the positive *z*-axis, where

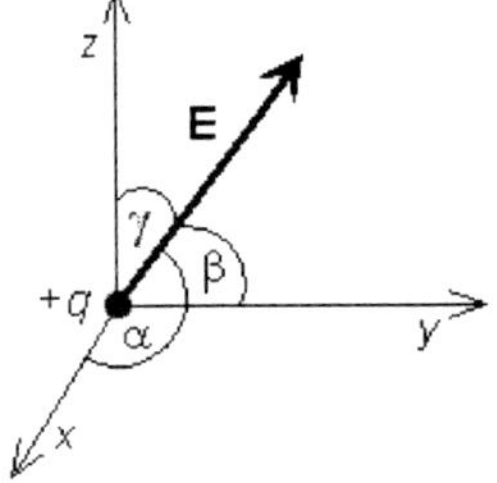

Fig. 3.14.

$$\gamma = \cos^{-1}\left(\pm\sqrt{1 - \cos^2\alpha - \cos^2\beta}\right)$$

$$= \cos^{-1}\left(\sqrt{1 - \cos^2 120° - \cos^2 45°}\right)$$

$$= \cos^{-1}(0.5)$$

$= 60°$, as the component of **E** along the *z*-axis is positive.

1(b) Solve for the electric field vector **E**, where

$$\mathbf{E} = |\,\mathbf{E}\,|\,(\cos\alpha\,\mathbf{i} + \cos\beta\,\mathbf{j} + \cos\gamma\,\mathbf{k})$$

$$= 200\,(\cos 120^\circ\,\mathbf{i} + \cos 45^\circ\,\mathbf{j} + \cos 60^\circ\,\mathbf{k})$$

$$= -100\,\mathbf{i} + 141.4214\,\mathbf{j} + 100\,\mathbf{k}\ \text{N/C}.$$

1(c) The force **F** on the electric charge is given by

$$\mathbf{F} = q\mathbf{E} = 6.75 \times 10^{-6}\,(-100\,\mathbf{i} + 141.4214\,\mathbf{j} + 100\,\mathbf{k})$$

$$= 10^{-4}\,(-6.75\,\mathbf{i} + 9.546\,\mathbf{j} + 6.75\,\mathbf{k})\ \text{N}.$$

Solve for the magnitude of **F**, | **F** |, given by

$$|\,\mathbf{F}\,| = \sqrt{(-6.75\times10^{-4})^2 + (9.546\times10^{-4})^2 + (6.75\times10^{-4})^2}$$

$$= 1.35 \times 10^{-3}\ \text{N}.$$

Vector Algebra Tools Solutions

Using the 3D Vector Components Tool: Enter the given magnitude for vector **E**, | **E** |, in the Magnitude of vector **a**, | **a** | user input field. Next enter the given angles α and β in the Angle α and Angle β user input fields, respectively. Then click the yellow Vector **a** and Angle γ button. The answers will be displayed in the output fields as shown.

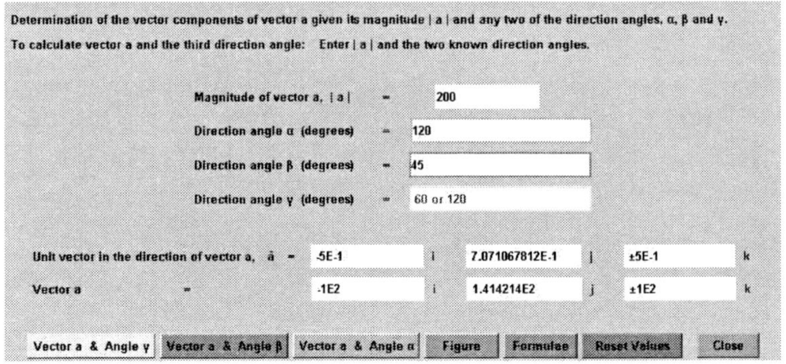

Solution: 1(a) $\gamma = 60^\circ$ and 1(b) $\mathbf{E} = -100\,\mathbf{i} + 141.4214\,\mathbf{j} + 100\,\mathbf{k}$ N/C.

Using the Vector Arithmetic Tool: Enter the calculated components of vector **E** in the Vector **a** user input fields and 0 for each of the components in the Vector **b** user input fields, together with the value of 6.75×10^{-6} for the Scalar multiplier, p user input field. Then click the Add **a** + **b** button. The answers will be displayed in the output fields as shown.

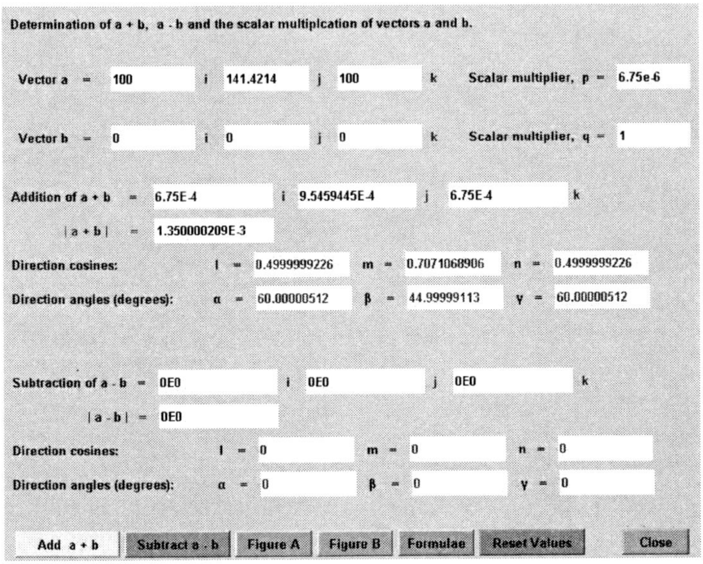

Determination of a + b, a - b and the scalar multiplcation of vectors a and b.

Vector a =	100	i	141.4214	j	100	k	Scalar multiplier, p = 6.75e-6
Vector b =	0	i	0	j	0	k	Scalar multiplier, q = 1
Addition of a + b =	6.75E-4	i	9.5459445E-4	j	6.75E-4	k	
| a + b | =	1.350000209E-3						
Direction cosines:	l =	0.4999999226	m =	0.7071068906	n =	0.4999999226	
Direction angles (degrees):	α =	60.00000512	β =	44.99999113	γ =	60.00000512	
Subtraction of a - b =	0E0	i	0E0	j	0E0	k	
| a - b | =	0E0						
Direction cosines:	l =	0	m =	0	n =	0	
Direction angles (degrees):	α =	0	β =	0	γ =	0	

Add a + b | Subtract a - b | Figure A | Figure B | Formulae | Reset Values | Close

Solution: 1(c) The magnitude of the force is $|\, \mathbf{F} \,| = 1.35 \times 10^{-3}$ N.

Answers: 1(a) $\gamma = 60°$, 1(b) $\mathbf{E} = -100\, \mathbf{i} + 141.4214\, \mathbf{j} + 100\, \mathbf{k}$ N/C and 1(c) $|\, \mathbf{F} \,| = 1.35 \times 10^{-3}$ N.

Question 2. A force **F** is applied to a particle P that is a distance of $|\, \mathbf{r} \,| = 0.25$ m from the origin O, as shown in figure 3.15. If $\mathbf{F} = 2\, \mathbf{i} + 3\, \mathbf{j} - 3\, \mathbf{k}$ N, determine the magnitude of the instantaneous torque $\boldsymbol{\tau}$ acting on the particle about the origin.

Worked Solution

We have $\ell = \cos \alpha = \cos 50° = 0.6428$, $m = \cos \beta$ and $n = \cos \gamma = \cos 60° = 0.5$. Solve for $\cos \beta$, where

$$\cos \beta = \pm \sqrt{1 - \cos^2 \alpha - \cos^2 \gamma}\,.$$

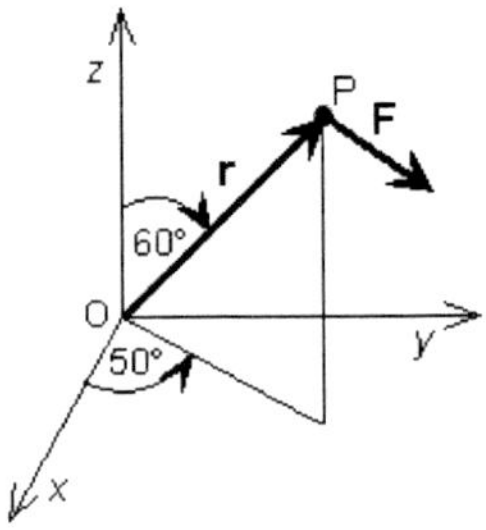

Fig. 3.15.

$$\cos\beta = \sqrt{1 - \cos^2 50^\circ - \cos^2 60^\circ}$$

$= 0.5804$, as the component of **r** along the y-axis is positive

Then solve for the displacement vector **r** of the particle P from the origin, that is on the line of action of the force, where

$$\mathbf{r} = |\mathbf{r}|(\cos\alpha\,\mathbf{i} + \cos\beta\,\mathbf{j} + \cos\gamma\,\mathbf{k})$$

$$= 0.25\,(0.6428\,\mathbf{i} + 0.5804\,\mathbf{j} + 0.5\,\mathbf{k})$$

$$= 0.1607\,\mathbf{i} + 0.1451\,\mathbf{j} + 0.125\,\mathbf{k}\text{ m.}$$

The instantaneous torque acting on the particle is given by $\boldsymbol{\tau} = \mathbf{r} \times \mathbf{F}$, where

$$\mathbf{r} \times \mathbf{F} = \begin{vmatrix} \mathbf{i} & \mathbf{j} & \mathbf{k} \\ a_x & a_y & a_z \\ b_x & b_y & b_z \end{vmatrix} = \begin{vmatrix} \mathbf{i} & \mathbf{j} & \mathbf{k} \\ 0.1607 & 0.1451 & 0.125 \\ 2 & 3 & -3 \end{vmatrix}$$

$$= -0.8103\,\mathbf{i} + 0.7321\,\mathbf{j} + 0.1919\,\mathbf{k}\text{ N m.}$$

Solve for the magnitude of $\boldsymbol{\tau}$, $|\boldsymbol{\tau}|$, given by

$$|\boldsymbol{\tau}| = \sqrt{(-0.8103)^2 + 0.7321^2 + 0.1919^2} = 1.09\text{ N m.}$$

Note that the moment arm of the force, which is the perpendicular distance d from the moment center at the origin to the line of action of the force, is given by the expression $d = \dfrac{|\boldsymbol{\tau}|}{|\mathbf{F}|}$.

Vector Algebra Tools Solution

Using the 3D Vector Components Tool: Enter the given magnitude for vector **r**, | **r** |, in the Magnitude of vector **a**, | **a** | user input field. Next enter the given angles α and γ in the Angle α and Angle γ user input fields, respectively. Then click the blue Vector **a** and Angle β button. The answers will be displayed in the answers will be displayed in the output fields as shown.

Determination of the vector components of vector a given its magnitude | a | and any two of the direction angles, α, β and γ.

To calculate vector a and the third direction angle: Enter | a | and the two known direction angles.

Magnitude of vector a, | a | = 0.25

Direction angle α (degrees) = 50

Direction angle β (degrees) = 54.52374743 or 125.47625257

Direction angle γ (degrees) = 60

Unit vector in the direction of vector a, â = 6.427876097E-1 i ±5.803654787E-1 j 5E-1 k

Vector a = 1.606969E-1 i ±1.450914E-1 j 1.25E-1 k

Vector a & Angle γ | Vector a & Angle β | Vector a & Angle α | Figure | Formulae | Reset Values | Close

Solution: cos β = 0.5804 is the component of the unit vector of **r** in the direction of the positive *y*-axis and **r** = 0.1607 **i** + 0.1451 **j** + 0.125 **k** m.

Using the Vector Product Tool: Enter the given components of vector **r** in the Vector **a** user input fields, and enter the given components of vector **F** in the Vector **b** user input fields. Then click the Vector Product **a × b** button. The answers will be displayed in the output fields as shown.

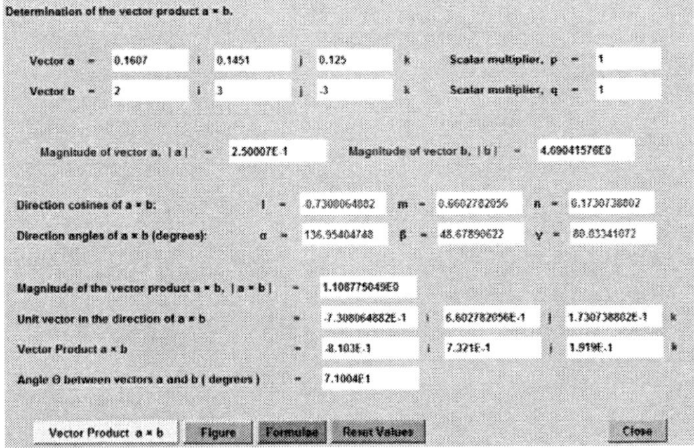

Solution: Magnitude of the instantaneous torque, | **τ** | = 1.09 N m.

Answer: The magnitude of the instantaneous torque on the particle about the origin is | **τ** | = 1.09 N m.

3.5 NOTES:

CHAPTER 4

SCALAR PRODUCT

In this chapter the scalar product, which is the product of two vectors resulting in a scalar quantity, is presented. In particular, scalar product definitions, properties of scalar product algebra, problem solving applications and worked examples are included. Also demonstrated is the use of the Scalar Product software tool in solving the example problems.

4.1 DEFINITION OF THE SCALAR PRODUCT

The **scalar product** or **dot product** is a scalar that is the product of two vectors. Specifically, the scalar product of any two vectors **a** and **b**, whose directions are at an angle of θ to each other, is defined by a scalar quantity that is the product of the magnitudes of the two vectors and the cosine of the angle θ between them, and is given by

$$\mathbf{a} \bullet \mathbf{b} = |\,\mathbf{a}\,|\,|\,\mathbf{b}\,| \cos\theta, \text{ where } 0° \le \theta \le 180°.$$

Geometrically, the scalar product **a** • **b** can be interpreted as the product of the **scalar projection** of vector **a** onto any nonzero vector **b** that is given by the length OP = | **a** | cos θ and the magnitude of vector **b**, | **b** |. See figure 4.1. Note that | **a** | cos θ can also be written as $\mathbf{a} \bullet \hat{\mathbf{b}}$, where $\hat{\mathbf{b}}$ is the unit vector in the direction of **b**, and is positive, zero or negative in value depending on the angle θ, where 0° ≤ θ ≤ 180°.

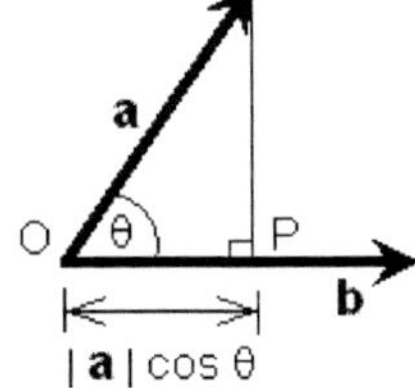

Fig. 4.1. The length OP = | **a** | cos θ is the scalar projection of **a** onto **b**, where **a** • **b** = | **a** | | **b** | cos θ.

4.2 DETERMINATION OF SCALAR PRODUCT QUANTITIES

4.2.1 Scalar product

If the vectors **a** and **b** are represented in their component form in a three-dimensional Cartesian coordinate system, where **a** = a_x **i** + a_y **j** +

a_z **k** and **b** = b_x **i** + b_y **j** + b_z **k**, then the scalar product of vectors **a** and **b** can be determined algebraically as follows:

$$\mathbf{a} \bullet \mathbf{b} = a_x\, b_x + a_y\, b_y + a_z\, b_z\,,$$

which is the sum of the products of the components for each vector along the **i**, **j** and **k** unit vectors, respectively.

4.2.2 Scalar projection

The **scalar projection** of vector **a** *along, onto* or *in the direction* of any nonzero vector **b** is a real number and is expressed as

$$\mathbf{a} \bullet \hat{\mathbf{b}} = |\,\mathbf{a}\,| \cos\theta = |\,\mathbf{a}\,| \left(\frac{\mathbf{a} \bullet \mathbf{b}}{|\mathbf{a}|\,|\,\mathbf{b}\,|} \right)$$

$$= \frac{\mathbf{a} \bullet \mathbf{b}}{|\,\mathbf{b}\,|} = \frac{a_x b_x + a_y b_y + a_z b_z}{\sqrt{(b_x^2 + b_y^2 + b_z^2)}}\,,$$

where $\hat{\mathbf{b}}$ is the unit vector that is in the direction of vector **b**. The real number that results from the scalar projection is called the **component** or **resolute** of vector **a** in the direction of vector **b**. Recall that the components of $\hat{\mathbf{b}}$ are the direction cosines of vector **b** and therefore we can also express the scalar projection as

$$\mathbf{a} \bullet \hat{\mathbf{b}} = a_x \cos\alpha + a_y \cos\beta + a_z \cos\gamma,$$

where the angles α, β and γ are the direction angles that vector **b** makes with the positive *x*, *y*, and *z*-axes, respectively.

4.2.3 Vector projection

The **vector projection** of vector **a** *along, onto* or *in the direction* of any nonzero vector **b** is a vector quantity and is expressed as

$$(|\,\mathbf{a}\,| \cos\theta)\,\hat{\mathbf{b}} = \left(\frac{\mathbf{a} \bullet \mathbf{b}}{|\,\mathbf{b}\,|} \right) \hat{\mathbf{b}} = \left(\frac{\mathbf{a} \bullet \mathbf{b}}{|\,\mathbf{b}\,|} \right) \frac{\mathbf{b}}{|\,\mathbf{b}\,|} = \left(\frac{\mathbf{a} \bullet \mathbf{b}}{|\,\mathbf{b}\,|^2} \right) \mathbf{b}$$

$$= \left(\frac{a_x b_x + a_y b_y + a_z b_z}{b_x^2 + b_y^2 + b_z^2} \right) (b_x \, \mathbf{i} + b_y \, \mathbf{j} + b_z \, \mathbf{k}),$$

where **b** / | **b** | is equal to the unit vector $\hat{\mathbf{b}}$ that is in the direction of vector **b**. Clearly, the vector projection of vector **a** onto vector **b** is simply the product of the scalar projection of vector **a** onto vector **b** and the unit vector $\hat{\mathbf{b}}$. Hence, vector **a** is said to be resolved into its **vector components** or **vector resolutes** along vector **b**.

4.2.4 The angle between two vectors

The angle θ between the directions of vectors **a** and **b** is given by

$$\theta = \cos^{-1} \left(\frac{\mathbf{a} \bullet \mathbf{b}}{|\mathbf{a}||\mathbf{b}|} \right)$$

$$= \cos^{-1} \left(\frac{a_x b_x + a_y b_y + a_z b_z}{\sqrt{(a_x^2 + a_y^2 + a_z^2)(b_x^2 + b_y^2 + b_z^2)}} \right),$$

where $0° \leq \theta \leq 180°$. If vectors **a** and **b** are **parallel**, pointing in the same direction, then $\theta = \cos^{-1}(1) = 0°$. However, if vectors **a** and **b** are **antiparallel**, pointing in opposite directions, then $\theta = \cos^{-1}(-1) = 180°$. In contrast, if vectors **a** and **b** are **orthogonal**, i.e., perpendicular to each other, then $\theta = \cos^{-1}(0) = 90°$.

4.3 LAWS AND PROPERTIES OF SCALAR PRODUCT ALGEBRA

Important laws and properties in scalar product algebra for the vectors **a**, **b** and **c** and the scalar quantities p and q are summarized below.

$\mathbf{a} \bullet \mathbf{b} = \mathbf{b} \bullet \mathbf{a}$ Commutative law

$\mathbf{a} \bullet (\mathbf{b} + \mathbf{c}) = \mathbf{a} \bullet \mathbf{b} + \mathbf{a} \bullet \mathbf{c}$ Distributive law

$p\mathbf{a} \bullet q\mathbf{b} = pq(\mathbf{a} \bullet \mathbf{b})$

$p(\mathbf{a} \bullet \mathbf{b}) = (p\mathbf{a}) \bullet \mathbf{b} = \mathbf{a} \bullet (p\mathbf{b}) = (\mathbf{a} \bullet \mathbf{b})p$

These axioms of scalar product algebra apply only to nonzero vectors and when the scalar quantities are nonzero real numbers, thereby permitting the manipulation of expressions involving scalar products in the same manner as ordinary algebraic equations.

4.4 IMPORTANT RESULTS INVOLVING SCALAR PRODUCTS

Important results and operations involving scalar products are outlined as follows:

- A scalar product quantity is always (a) positive when the angle θ between the two vectors, drawn tail-to-tail, is in the range $0° \le \theta < 90°$, (b) zero when $\theta = 90°$ and (c) negative when θ is in the range $90° < \theta \le 180°$. See figure 4.2.

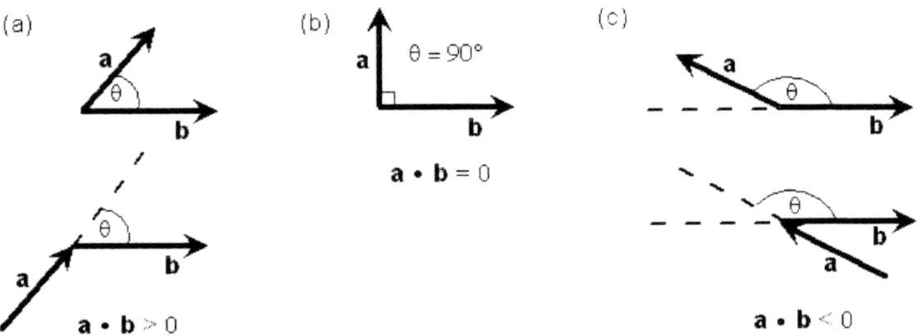

Fig. 4.2. Scalar products of the vectors **a** and **b**, where the angle θ between the directions of the vectors **a** and **b** are in the range (a) $0° \le \theta < 90°$, (b) $\theta = 90°$ and (c) $90° < \theta \le 180°$.

- If vectors **a** and **b** are equal, where **b** = **a**, then

$$\mathbf{a} \bullet \mathbf{b} = \mathbf{a} \bullet \mathbf{a} = \mathbf{a}^2 = |\,\mathbf{a}\,|^2 = a^2.$$

  Furthermore, if vector **a** is written in component form as $\mathbf{a} = a_x\,\mathbf{i} + a_y\,\mathbf{j} + a_z\,\mathbf{k}$, then

$$\mathbf{a} \bullet \mathbf{a} = a_x^{\,2} + a_y^{\,2} + a_z^{\,2}.$$

  Note that the square root of **a** • **a** is the magnitude of vector **a**.

- If vectors **a** and **b** are **parallel**, pointing in the same direction, then the scalar product $\mathbf{a} \bullet \mathbf{b} = ab$ (or $|\,\mathbf{a}\,|\,|\,\mathbf{b}\,|$). However, if vectors **a** and **b** are **antiparallel**, pointing in opposite directions, then $\mathbf{a} \bullet \mathbf{b} = -ab$ (or $-|\,\mathbf{a}\,|\,|\,\mathbf{b}\,|$).

- If **i**, **j** and **k** are the unit vectors that are along the mutually perpendicular positive *x*, *y* and *z*-axes of a three-dimensional Cartesian coordinate system, respectively, then the scalar product of two unit vectors that are parallel and in the same direction can be expressed as

$$\mathbf{i} \bullet \mathbf{i} = \mathbf{i}^2 = 1, \ \mathbf{j} \bullet \mathbf{j} = \mathbf{j}^2 = 1 \ \text{and} \ \mathbf{k} \bullet \mathbf{k} = \mathbf{k}^2 = 1.$$

- If either one or both of the vectors **a** and **b** are zero vectors, then $\mathbf{a} \bullet \mathbf{b} = 0$. However, if the vectors **a** and **b** are nonzero vectors and are **orthogonal**, then $\theta = 90°$, $\cos\theta = 0$ and the scalar product of $\mathbf{a} \bullet \mathbf{b} = 0$. Note that this **orthogonal condition** is commonly used in vector algebra to test if two nonzero vectors are orthogonal to each other. For example, if vectors **a** and **b** are written in component form as $\mathbf{a} = a_x\,\mathbf{i} + a_y\,\mathbf{j} + a_z\,\mathbf{k}$ and $\mathbf{b} = b_x\,\mathbf{i} + b_y\,\mathbf{j} + b_z\,\mathbf{k}$ and are orthogonal, then

$$a_x\,b_x + a_y\,b_y + a_z\,b_z = 0.$$

- If **i**, **j** and **k** are the unit vectors that are along the respective mutually perpendicular positive x, y and z-axes of a three-dimensional Cartesian coordinate system, then, based upon the orthogonal condition,

$$\mathbf{j} \bullet \mathbf{k} = \mathbf{k} \bullet \mathbf{i} = \mathbf{i} \bullet \mathbf{j} = 0.$$

- If vector **a** is written in component form as $\mathbf{a} = a_x\,\mathbf{i} + a_y\,\mathbf{j} + a_z\,\mathbf{k}$, then the scalar projections of vector **a** onto the corresponding unit vectors **i**, **j** and **k** are expressed as follows:

$$\begin{aligned}
\mathbf{a} \bullet \mathbf{i} &= a_x \times 1 + a_y \times 0 + a_z \times 0 = a_x, \\
\mathbf{a} \bullet \mathbf{j} &= a_x \times 0 + a_y \times 1 + a_z \times 0 = a_y, \\
\mathbf{a} \bullet \mathbf{k} &= a_x \times 0 + a_y \times 0 + a_z \times 1 = a_z,
\end{aligned}$$

where a_x, a_y and a_z are the x, y, z components of vector **a** along the positive x, y and z-axes, respectively. Note that the scalar projections of vector **a** onto the corresponding unit vectors **i**, **j** and **k** can also be expressed as

$$\mathbf{a} \bullet \mathbf{i} = |\mathbf{a}| \cos\alpha, \ \mathbf{a} \bullet \mathbf{j} = |\mathbf{a}| \cos\beta \ \text{and} \ \mathbf{a} \bullet \mathbf{k} = |\mathbf{a}| \cos\gamma,$$

where $|\mathbf{a}|$ is the magnitude of vector **a** and α, β and γ are the direction angles of vector **a** measured from the positive x, y and z-axes, respectively.

- For vectors **a** and **b**, the sum of the square of the magnitudes of their scalar and vector products can be expressed as

$$|\mathbf{a} \bullet \mathbf{b}|^2 + |\mathbf{a} \times \mathbf{b}|^2 = |\mathbf{a}|^2 |\mathbf{b}|^2 \cos^2\theta + |\mathbf{a}|^2 |\mathbf{b}|^2 \sin^2\theta$$

$$= |\mathbf{a}|^2 |\mathbf{b}|^2 (\cos^2\theta + \sin^2\theta)$$

$$= |\mathbf{a}|^2 + |\mathbf{b}|^2 = a^2 + b^2.$$

4.5 PROBLEM SOLVING APPLICATIONS OF SCALAR PRODUCTS

The following examples demonstrate the numerous problem solving applications of scalar products.

- Determination of the kinetic energy of a particle of mass m can be expressed as $\frac{1}{2}m(\mathbf{v} \bullet \mathbf{v})$ or $\frac{1}{2}m\mathbf{v}^2$, where **v** is the instantaneous velocity of the particle.

- Determination of the work W done by a constant force **F** on an object that moves along the displacement vector **s**, where W is given by $W = \mathbf{F} \bullet \mathbf{s}$. Since the component of **F** in the direction of the displacement vector **s** is $|\mathbf{F}| \cos\theta$, the work done by **F** in moving the object is given by $W = (|\mathbf{F}| \cos\theta) |\mathbf{s}|$, where W is positive in value since work is done *by* the system. See figure 4.3 (a). In contrast, as shown in figure 4.3 (b), the work done by the force of friction $\mathbf{F}_f$ that opposes the direction of motion is given by $\mathbf{F}_f \bullet \mathbf{s} = |\mathbf{F}_f| |\mathbf{s}| \cos 180° = -|\mathbf{F}_f| |\mathbf{s}|$ and W is negative in value, since work is being done *on* the system.

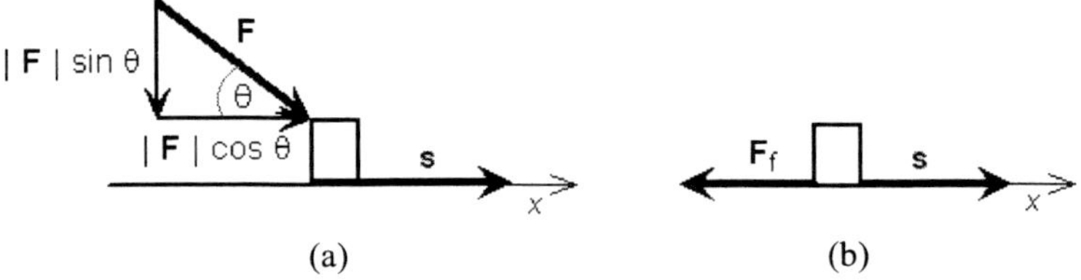

Fig. 4.3. (a) The applied force **F** does work in moving the object along the displacement vector **s**, such that work is done *by* the system and is given by $\mathbf{F} \bullet \mathbf{s} = (|\mathbf{F}| \cos\theta) |\mathbf{s}|$. (b) In contrast, the force of friction $\mathbf{F}_f$ opposes the direction of motion, such that work is done *on* the system and is given by $\mathbf{F}_f \bullet \mathbf{s} = -|\mathbf{F}_f| |\mathbf{s}|$.

- Determination of the total work done by the net external force $\mathbf{F}_{net}$ acting on a particle of mass m that undergoes a displacement $\mathbf{s}$ and is given by the expression

$$W = \mathbf{F}_{net} \bullet \mathbf{s}$$

$$= \Delta K = \tfrac{1}{2}m(\mathbf{v}_f \bullet \mathbf{v}_f) - \tfrac{1}{2}m(\mathbf{v}_i \bullet \mathbf{v}_i) = \tfrac{1}{2}m(\mathbf{v}_f^2 - \mathbf{v}_i^2),$$

where $\mathbf{v}_i$ and $\mathbf{v}_f$ are the initial and final velocities of the particle, respectively. Thus, the total work done by $\mathbf{F}_{net}$ is equal to the change in the kinetic energy, ΔK, of the displaced particle. This result is commonly known as the **work-energy theorem**.

- Determination of the change in potential energy, ΔU, when a **conservative force** $\mathbf{F}_c$ on a particle that undergoes a displacement $\mathbf{s}$, and is given by the expression

$$\Delta U = -\mathbf{F}_c \bullet \mathbf{s}.$$

For example, when a particle falls in a gravitational field, the work done by the conservative gravitational force does positive work on the particle and the particle's potential energy decreases. Conversely, when the particle is thrown up, the gravitational force does negative work on the particle and the potential energy of the particle increases. From the work-energy theorem for a conservative force we can write

$$\Delta K = -\Delta U.$$

This expression is known as the **law of conservation of mechanical energy** and implies that the sum of the kinetic energy and potential energy of the particle, i.e., the total mechanical energy is constant.

- Determination of the instantaneous power P or the time rate at which the work is done by a constant net force $\mathbf{F}_{net}$ on a particle is given by $P = \mathbf{F}_{net} \bullet \mathbf{v}$, where $\mathbf{v}$ is the instantaneous velocity of the particle.

- Determination of the electric current I at a point in a conductor that corresponds to the flow of electric charge through a given area A, which can be expressed as $I = \mathbf{J} \bullet \mathbf{A}$, where $\mathbf{J}$ is the electric current density and $\mathbf{A}$ ($= A\,\hat{\mathbf{n}}$) is the directed area vector in the direction of unit vector $\hat{\mathbf{n}}$.

- Determination of the electric flux Φ_E through a surface of fixed area A in a uniform external electric field **E**, which can be expressed as $\Phi_E = \mathbf{E} \bullet \mathbf{A}$, where $\mathbf{A}$ $(= A\,\hat{\mathbf{n}})$ is the directed area vector normal to the surface and, by convention, always points outward, as shown in figure 4.4.

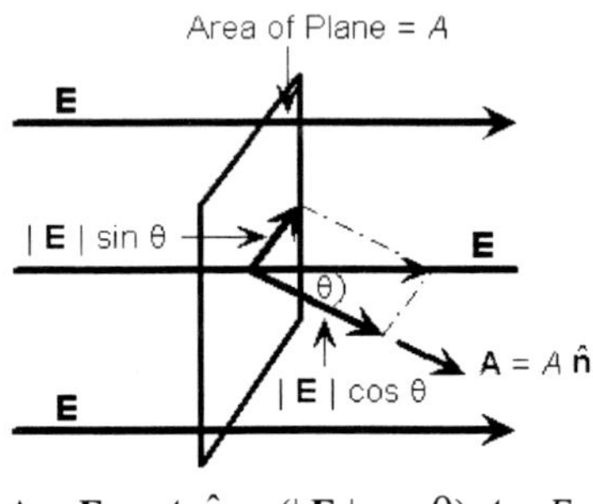

$$\Phi_E = \mathbf{E} \bullet \mathbf{A} = \mathbf{E} \bullet A\;\hat{\mathbf{n}} = (|\,\mathbf{E}\,|\cos\theta)\,A = E\,A\cos\theta$$

Fig. 4.4. Electric field lines that pass through a planar surface of area A tilted at an angle θ away from **E**.

For a point charge q located outside any closed surface, for example a cylinder, the net electric flux through the closed surface is zero. In this case, the number of electric field lines entering the surface equals the number leaving the surface. See figure 4.5. Specifically, the sign of the electric flux Φ_E is negative for the electric field lines that enter the surface and is positive for the electric field lines that leave the surface.

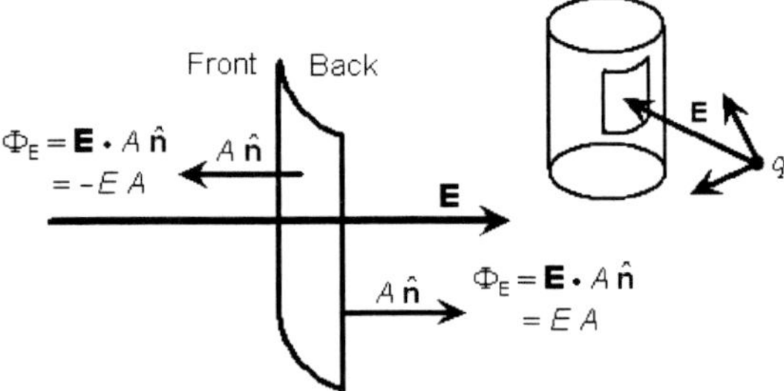

Electric field **E** entering an area element A of the front surface of a cylinder.

Electric field **E** leaving an area element A of the front surface of a cylinder.

Fig. 4.5. Net electric flux through an element of area of a closed surface of a cylinder is zero.

Note that, for any arbitrary closed or gaussian surface that surrounds a net electric charge, Q, Gauss' law states that the net electric flux, Φ_C, through the closed surface is equal to the net charge Q inside the surface divided by ε_o, where ε_o is the permittivity of free space, such that

$$\Phi_C = \frac{Q}{\varepsilon_o}.$$

- The magnetic flux associated with a magnetic field is defined in a manner similar to that used to define the electric flux. For example, the magnetic flux Φ_B through a surface of fixed area A in a uniform external magnetic field **B** can be expressed as $\Phi_B = \mathbf{B} \bullet \mathbf{A}$, where **A** ($= A\,\hat{\mathbf{n}}$) is the directed area vector normal to the surface, and, by convention, always points outward. Note that since magnetic fields are continuous and form closed loops in nature, e.g., the magnetic field lines of a bar magnet, Gauss' law in magnetism states that the net magnetic flux through any closed surface is *always* zero.

- Determination of the potential energy U_E of an electric dipole defined as $U_E = -\mathbf{p} \bullet \mathbf{E}$, where **p** is the electric dipole moment and **E** is the uniform external electric field. This type of calculation is frequently used in physical chemistry to determine the potential energy of a polar molecule, e.g., HCl.

- Determination of the potential energy U_B of a magnetic dipole in a uniform external electric field defined as $U_B = -\boldsymbol{\mu} \bullet \mathbf{B}$, where $\boldsymbol{\mu}$ is the magnetic dipole moment and **B** is the magnetic field.

- The **law of cosines** for any plane triangle ABC where the vectors **a**, **b** and **c** are the sides of the triangle, as shown in figure 4.6, can be expressed as follows:

$$\mathbf{a} \bullet \mathbf{a} = \mathbf{a}^2 = (\mathbf{c} - \mathbf{b})^2 = (\mathbf{c} \bullet \mathbf{c}) + (\mathbf{b} \bullet \mathbf{b}) - 2(\mathbf{c} \bullet \mathbf{b})$$

$$= |\,\mathbf{c}\,|^2 + |\,\mathbf{b}\,|^2 - 2\,|\,\mathbf{c}\,|\,|\,\mathbf{b}\,|\cos\theta.$$

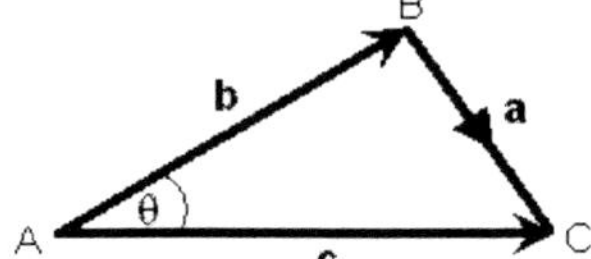

Fig. 4.6. The plane triangle ABC with sides **a**, **b** and **c**, where θ is the angle between vectors **b** and **c**.

- Determination of the angle θ between two straight lines that are expressed by their vector equations $\mathbf{r} = \mathbf{a}_1 + \lambda_1\mathbf{b}_1$ and $\mathbf{r} = \mathbf{a}_2 + \mu\mathbf{b}_2$ is given by

$$\theta = \cos^{-1}\left(\frac{\mathbf{b}_1 \bullet \mathbf{b}_2}{|\mathbf{b}_1|\,|\mathbf{b}_2|}\right).$$

  This expression also applies in the case of a pair of skew or a pair of intersecting lines. Note that if the two lines are perpendicular, then $\mathbf{b}_1 \bullet \mathbf{b}_2 = 0$; if the lines are parallel, then $\mathbf{b}_1$ and $\mathbf{b}_2$ are collinear vectors, such that $\mathbf{b}_1 = \lambda\mathbf{b}_2$, where λ is a nonzero real number scalar.

- Determination of the perpendicular distance $|\mathbf{a}_\perp|$ from a point P that has a position vector **p** from the origin to a straight line, defined by the equation $\mathbf{r} = \mathbf{a} + \lambda\mathbf{b}$, and is given by the expression

$$|\mathbf{a}_\perp| = [(\mathbf{p} - \mathbf{a})^2 - \frac{[(\mathbf{p} - \mathbf{a}) \bullet \mathbf{b}]^2}{|\mathbf{b}|^2}]^{1/2},$$

  where **a** is the position vector of a point on the line. Note that the scalar projection of the vector $\mathbf{p} - \mathbf{a}$ onto the line is given by $\dfrac{(\mathbf{p} - \mathbf{a}) \bullet \mathbf{b}}{|\mathbf{b}|}$.

- Determination of the vector equation of a plane that is expressed in terms of the position vector of a point in the plane and a vector normal to the plane. See figure 4.7. For example, if the given point in the plane is A, with a position vector **a** from the origin at point O, and **n** is the vector normal to the plane, whose direction is specified by the unit vector $\hat{\mathbf{n}}$, then **n** is also perpendicular to the vector $\mathbf{p} = \mathbf{r} - \mathbf{a}$ that lies in the plane. Therefore, the scalar product form of the vector equation of the plane can be written as

$$\mathbf{p} \bullet \mathbf{n} = (\mathbf{r} - \mathbf{a}) \bullet \mathbf{n} = 0 \text{ or } \mathbf{r} \bullet \mathbf{n} = \mathbf{a} \bullet \mathbf{n},$$

  where **r** is the position vector of any general point P in the plane. Hence knowing **a** and **n** we can evaluate $\mathbf{a} \bullet \mathbf{n}$, which is equal to

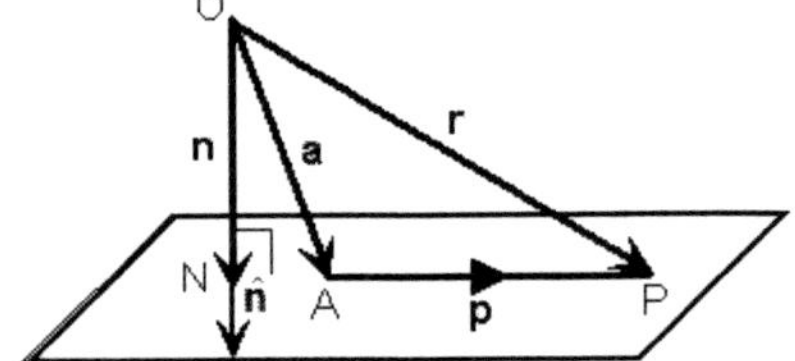

Fig. 4.7. Vector equation of a plane $\mathbf{r} \bullet \mathbf{n} = \mathbf{a} \bullet \mathbf{n}$ expressed in terms of the position vector of a point A in the plane and a vector **n** normal to the plane.

the scalar d, and the scalar product form of the vector equation of the plane can be written as

$$\mathbf{r} \bullet \mathbf{n} = d.$$

In the case where $\mathbf{n} = \hat{\mathbf{n}}$, then $\mathbf{a} \bullet \hat{\mathbf{n}}$ is the projection of vector **a** onto vector **n** and is equal to p, which is the perpendicular distance of the plane from the origin. Thus, the normal or perpendicular form of the vector equation of the plane can be written as

$$\mathbf{r} \bullet \hat{\mathbf{n}} = p.$$

Note that p is positive, since the scalar product $\mathbf{s} \bullet \hat{\mathbf{n}}$ is positive and the angle between the two vectors is acute. Thus, it follows that the Cartesian form of the vector equation for any general point in the plane given by $\mathbf{r} = x\,\mathbf{i} + y\,\mathbf{j} + z\,\mathbf{k}$, can be expressed as

$$\begin{aligned}\mathbf{r} \bullet \hat{\mathbf{n}} &= (x\,\mathbf{i} + y\,\mathbf{j} + z\,\mathbf{k}) \bullet (\ell\,\mathbf{i} + m\,\mathbf{j} + n\,\mathbf{k}) \\ &= x\ell + ym + zn = p,\end{aligned}$$

where ℓ, m, n are the direction cosines of the vector normal **n** to the plane. Moreover, if $ax + by + cz = d$ represents the general form of the Cartesian equation of a plane, then the scalar product form of vector equation of the same plane can be expressed as

$$\mathbf{r} \bullet \mathbf{n} = \mathbf{r} \bullet (a\,\mathbf{i} + b\,\mathbf{j} + c\,\mathbf{k}) = d,$$

where a, b and c are the components of the vector normal **n** to the plane and the perpendicular distance p of the plane from the origin is given by $p = \dfrac{d}{|\mathbf{n}|} = \dfrac{d}{\sqrt{a^2 + b^2 + c^2}}$.

- If the equations of two planes are expressed in their normal form, then the angle θ between the two planes is given by

$$\theta = \cos^{-1}(\hat{\mathbf{n}}_1 \bullet \hat{\mathbf{n}}_2),$$

where $\hat{\mathbf{n}}_1$ and $\hat{\mathbf{n}}_2$ are the unit vectors that are in the direction of the vectors normal to each plane.

4.6 WORKED EXAMPLES

The **Scalar Product Tool** can be used to determine the scalar product of any two vectors in component form in a Cartesian coordinate system. In addition, the Scalar Product Tool can be used to determine the scalar projection and vector projection of one vector onto another vector, and to calculate the angle between the two vectors. See Appendix B for information on how to enter values into any of the Vector Algebra Tools.

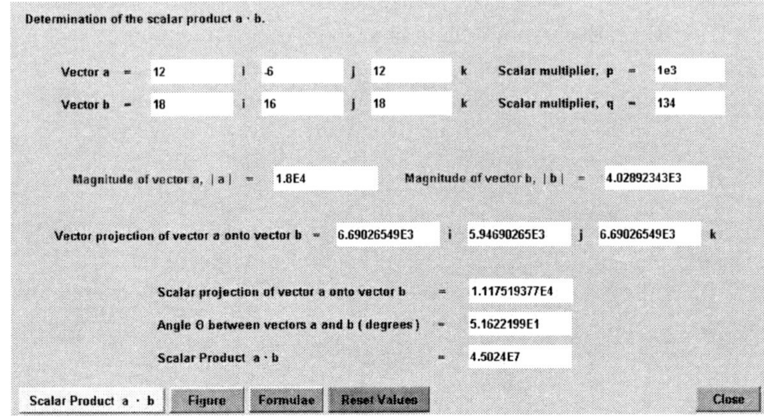

Fig. 4.8. The Scalar Product Tool from the Vector Algebra Tools Software Program.

Question 1. What is the geometrical interpretation of the scalar product $\mathbf{a} \bullet \mathbf{b}$, where $\mathbf{a} = -4\,\mathbf{i} + 7\,\mathbf{j} - 5\,\mathbf{k}$ and $\mathbf{b} = 10\,\mathbf{i} - 17.5\,\mathbf{j} + 12.5\,\mathbf{k}$? The unit of length is the meter.

Worked Solution

Solve for $\mathbf{a} \bullet \mathbf{b}$, where

$$\mathbf{a} \bullet \mathbf{b} = a_x\, b_x + a_y\, b_y + a_z\, b_z = -4(10) + 7(-17.5) - 5(12.5) = -225 \text{ m}.$$

Determine the angle θ between the directions of vectors **a** and **b**, where

$$\theta = \cos^{-1}\left(\frac{\mathbf{a} \bullet \mathbf{b}}{|\mathbf{a}|\,|\mathbf{b}|}\right) = \cos^{-1}\left(\frac{a_x b_x + a_y b_y + a_z b_z}{\sqrt{a_x^2 + a_y^2 + a_z^2}\sqrt{b_x^2 + b_y^2 + b_z^2}}\right)$$

$$= \cos^{-1}\left(\frac{-225}{\sqrt{(-4)^2 + 7^2 + (-5)^2}\sqrt{10^2 + (-17.5)^2 + 12.5^2}}\right)$$

$$= \cos^{-1}(-1.0) = 180°.$$

Since $\theta = 180°$, the vectors **a** and **b** are collinear and antiparallel. Therefore, $\mathbf{a} = \lambda\mathbf{b}$, where $\lambda = -0.4$.

Vector Algebra Tools Solution

Using the Scalar Product Tool: Enter the given components of vector **a** in the Vector **a** user input fields and the given components of vector **b** in the Vector **b** user input fields. Then click the Scalar Product **a • b** button. The answers will be displayed in the output fields as shown.

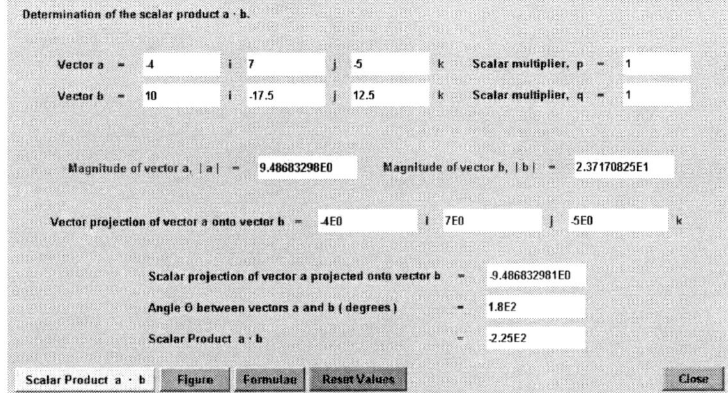

Solution: The angle θ between the directions of vectors **a** and **b** is 180°.

Answer: Since $\theta = 180°$, vectors **a** and **b** are collinear and antiparallel. Therefore, $\mathbf{a} = \lambda\mathbf{b}$, where $\lambda = -0.4$.

Question 2. Determine the component of the force vector $\mathbf{F} = 8\,\mathbf{i} - 5\,\mathbf{j} + 6\,\mathbf{k}$ N that is in the direction of the displacement vector $\mathbf{s} = 9\,\mathbf{i} + 3\,\mathbf{j} + 4\,\mathbf{k}$ m.

Worked Solution

The component or scalar projection of **F** in the direction of the displacement vector **s** is given by

$$\mathbf{F} \bullet \hat{\mathbf{s}} = \frac{\mathbf{F} \bullet \mathbf{s}}{|\,\mathbf{s}\,|}.$$

Determine $\mathbf{F} \bullet \mathbf{s}$, where

$$\mathbf{F} \bullet \mathbf{s} = a_x\,b_x + a_y\,b_y + a_z\,b_z = 8(9) + (-5)(3) + 6(4) = 81 \text{ N m}.$$

Then solve for $\dfrac{\mathbf{F} \bullet \mathbf{s}}{|\,\mathbf{s}\,|}$, where

$$\frac{\mathbf{F} \bullet \mathbf{s}}{|\,\mathbf{s}\,|} = \frac{81}{\sqrt{9^2 + 3^2 + 4^2}} = \frac{81}{10.2956} = 7.8674 \text{ N.}$$

Alternatively, we can express the component of **F** in the direction of **s** as

$$\mathbf{F} \bullet \hat{\mathbf{s}} = \mathbf{F} \bullet \frac{\mathbf{s}}{|\,\mathbf{s}\,|}$$

$$= (8\,\mathbf{i} - 5\,\mathbf{j} + 6\,\mathbf{k}) \bullet \left(\frac{9\,\mathbf{i} + 3\,\mathbf{j} + 4\,\mathbf{k}}{10.2956}\right)$$

$$= (8\,\mathbf{i} - 5\,\mathbf{j} + 6\,\mathbf{k}) \bullet (0.87415727\,\mathbf{i} + 0.29138575\,\mathbf{j} + 0.38851434\,\mathbf{k})$$

$$= 7.8674 \text{ N.}$$

Note that the work W done by the force in the direction of the displacement vector **s** is $W = (\mathbf{F} \bullet \hat{\mathbf{s}})\,|\,\mathbf{s}\,|$.

Vector Algebra Tools Solution

Using the Scalar Product Tool: Enter the given components of vector **F** in the Vector **a** user input fields and the given components of vector **s** in the Vector **b** user input fields. Then click the Scalar Product **F • s** button. The answers will be displayed in the output fields as shown.

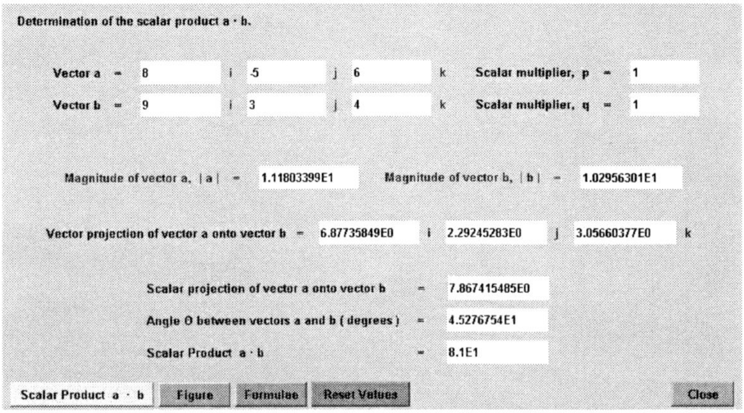

Determination of the scalar product a · b.

Vector a =	8	i	5	j	6	k	Scalar multiplier, p = 1
Vector b =	9	i	3	j	4	k	Scalar multiplier, q = 1

Magnitude of vector a, | a | = 1.11803399E1 Magnitude of vector b, | b | = 1.02956301E1

Vector projection of vector a onto vector b = 6.87735849E0 i 2.29245283E0 j 3.05660377E0 k

Scalar projection of vector a onto vector b = 7.867415485E0

Angle θ between vectors a and b (degrees) = 4.5276754E1

Scalar Product a · b = 8.1E1

Scalar Product a · b | Figure | Formulae | Reset Values | Close

Answer: The component of vector **F** in the direction of vector **s** = 7.8674 N. Check answer: | **F** | cos θ = 11.1803 cos 45.277° = 7.8674 N.

Question 3: In figure 4.9 the position vector **a** of point P from the origin O is resolved parallel, $\mathbf{a}_{\parallel}$, and perpendicular, $\mathbf{a}_{\perp}$, to the straight line OB that is in the direction of vector **b**, such that

$$\mathbf{a}_{\parallel} = (\mathbf{a} \bullet \hat{\mathbf{b}})\,\hat{\mathbf{b}} = \left(\frac{\mathbf{a} \bullet \mathbf{b}}{|\,\mathbf{b}\,|^2}\right)\mathbf{b} \text{ and } \mathbf{a}_{\perp} = \mathbf{a} - \mathbf{a}_{\parallel} = \mathbf{a} - \left(\frac{\mathbf{a} \bullet \mathbf{b}}{|\,\mathbf{b}\,|^2}\right)\mathbf{b}.$$

Calculate the vectors $\mathbf{a}_{\parallel}$ and $\mathbf{a}_{\perp}$ given that **a** = 4 **i** + 4 **j** + 3 **k** and **b** = 2 **i** + **j** + 2 **k**. *Hint*: Note that $\mathbf{a}_{\parallel}$ is the vector projection of **a** onto **b**. The unit of length is the meter.

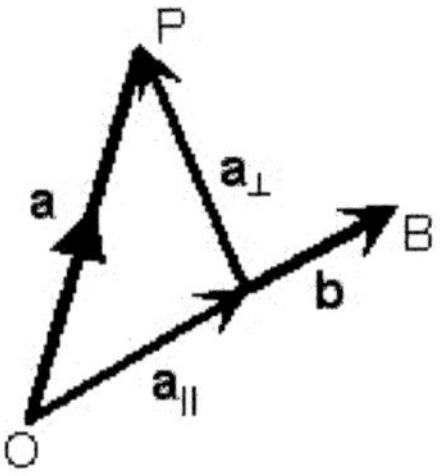

Fig. 4.9.

Worked Solutions

Solve for $\mathbf{a}_{\parallel}$, where

$$\mathbf{a}_{\parallel} = \left(\frac{\mathbf{a} \bullet \mathbf{b}}{|\,\mathbf{b}\,|^2}\right)\mathbf{b}$$

$$= \left(\frac{4(2) + 4(1) + 3(2)}{2^2 + 1^2 + 2^2} \right) (2\,\mathbf{i} + \mathbf{j} + 2\,\mathbf{k})$$

$$= 2(2\,\mathbf{i} + \mathbf{j} + 2\,\mathbf{k})$$

$$= 4\,\mathbf{i} + 2\,\mathbf{j} + 4\,\mathbf{k} \text{ m.}$$

Solve for $\mathbf{a}_{\perp}$, where

$$\mathbf{a}_{\perp} = \mathbf{a} - \left(\frac{\mathbf{a} \bullet \mathbf{b}}{|\mathbf{b}|^2} \right) \mathbf{b}$$

$$= (4\,\mathbf{i} + 4\,\mathbf{j} + 3\,\mathbf{k}) - \mathbf{a}_{\parallel}$$

$$= (4\,\mathbf{i} + 4\,\mathbf{j} + 3\,\mathbf{k}) - (4\,\mathbf{i} + 2\,\mathbf{j} + 4\,\mathbf{k})$$

$$= 2\,\mathbf{j} - \mathbf{k} \text{ m.}$$

Vector Algebra Tools Solutions

Using the Scalar Product Tool: Enter the given components of vector **a** in the Vector **a** user input fields and the given components of vector **b** in the Vector **b** user input fields. Then click the Scalar Product **a** • **b** button. The answers will be displayed in the output fields as shown.

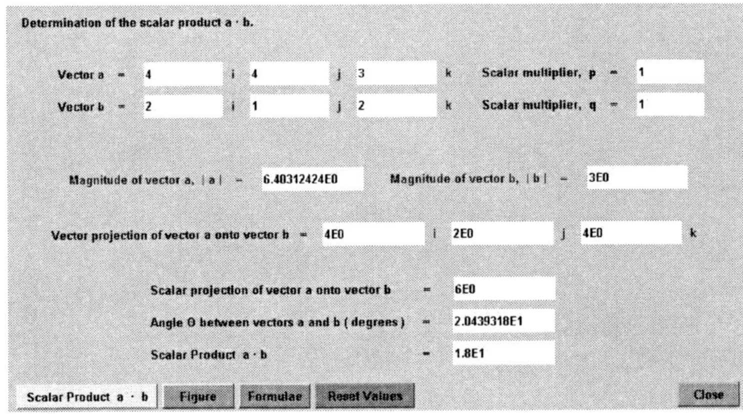

Solution: $\mathbf{a}_{\parallel} = 4\,\mathbf{i} + 2\,\mathbf{j} + 4\,\mathbf{k}$ m.

Using the Vector Arithmetic Tool: Enter the given components of vector **a** in the Vector **a** user input fields and enter the calculated components of vector $\mathbf{a}_{\parallel}$ in the Vector **b** user input fields. Then click the Subtract **a** − **b** button. The answers will be displayed in the output fields as shown.

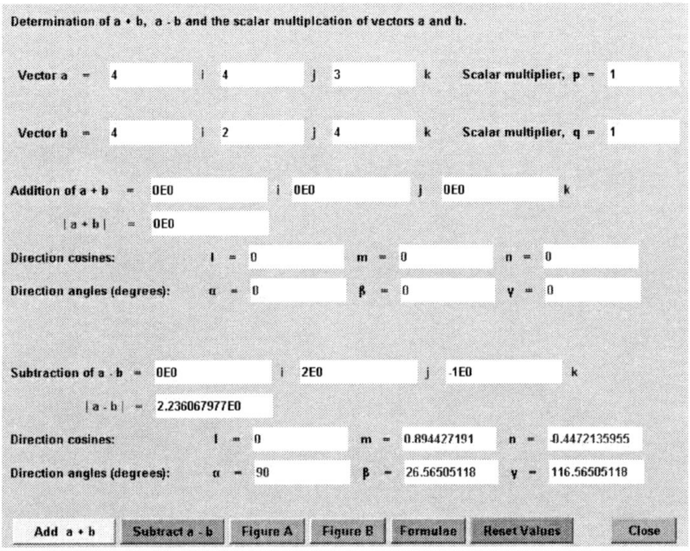

Determination of a + b, a - b and the scalar multiplcation of vectors a and b.

Vector a =	4	i	4	j	3	k	Scalar multiplier, p = 1
Vector b =	4	i	2	j	4	k	Scalar multiplier, q = 1
Addition of a + b =	0E0	i	0E0	j	0E0	k	
| a + b | =	0E0						
Direction cosines:	l = 0		m = 0		n = 0		
Direction angles (degrees):	α = 0		β = 0		γ = 0		
Subtraction of a - b =	0E0	i	2E0	j	-1E0	k	
| a - b | =	2.236067977E0						
Direction cosines:	l = 0		m = 0.894427191		n = -0.4472135955		
Direction angles (degrees):	α = 90		β = 26.56505118		γ = 116.56505118		

Add a + b | Subtract a - b | Figure A | Figure B | Formulae | Reset Values | Close

Solution: $\mathbf{a}_{\perp} = 2\,\mathbf{j} - \mathbf{k}$ m.

Answers: $\mathbf{a}_{\parallel} = 4\,\mathbf{i} + 2\,\mathbf{j} + 4\,\mathbf{k}$ m and $\mathbf{a}_{\perp} = 2\,\mathbf{j} - \mathbf{k}$ m. Note that $|\,\mathbf{a}_{\perp}\,|$ is the perpendicular distance from the point P defined by the position vector **a** from the origin to the straight line defined by the vector $\mathbf{r} = \lambda\mathbf{b}$ that passes through the origin.

Question 4: Forces of $|\,\mathbf{F}_1\,| = 42$ N in the direction of the vector $\mathbf{d}_1 = 3\,\mathbf{i} - 2\,\mathbf{j} + 6\,\mathbf{k}$ and $|\,\mathbf{F}_2\,| = 48$ N in the direction of the vector $\mathbf{d}_2 = 2\,\mathbf{i} - \mathbf{j} + 2\,\mathbf{k}$ act on an object of mass 8 kg. If the object is displaced from point S, having the position vector $\mathbf{s} = 2\,\mathbf{i} + 3\,\mathbf{j} - 4\,\mathbf{k}$ m, to the point P, having the position vector $\mathbf{p} = 11\,\mathbf{i} - 7\,\mathbf{j} - 6\,\mathbf{k}$ m, determine (a) the magnitude of the acceleration vector of the object and (b) the work done by the net external force during this displacement.

4(a) **Worked Solution**

The acceleration **a** of the object is given by the expression $\mathbf{a} = \dfrac{\mathbf{F}_{\text{net}}}{m}$, where $\mathbf{F}_{\text{net}}$ is the net external force applied to the object of mass m.

Solve for $\mathbf{F}_1$, where

$$\mathbf{F}_1 = |\,\mathbf{F}_1\,|\,\hat{\mathbf{d}}_1 = |\,\mathbf{F}_1\,|\,\frac{\mathbf{d}_1}{|\,\mathbf{d}_1\,|}$$

$$= 42\left(\frac{3\,\mathbf{i} - 2\,\mathbf{j} + 6\,\mathbf{k}}{7}\right)$$

$$= 18\,\mathbf{i} - 12\,\mathbf{j} + 36\,\mathbf{k}\text{ N.}$$

Solve for $\mathbf{F}_2$, where

$$\mathbf{F}_2 = |\,\mathbf{F}_2\,|\,\hat{\mathbf{d}}_2 = |\,\mathbf{F}_2\,|\,\frac{\mathbf{d}_2}{|\,\mathbf{d}_2\,|}$$

$$= 48\left(\frac{2\,\mathbf{i} - \mathbf{j} + 2\,\mathbf{k}}{3}\right)$$

$$= 32\,\mathbf{i} - 16\,\mathbf{j} + 32\,\mathbf{k}\text{ N.}$$

Determine $\mathbf{F}_{\text{net}}$, where

$$\mathbf{F}_{\text{net}} = \sum \mathbf{F} = \mathbf{F}_1 + \mathbf{F}_2$$

$$= (18\,\mathbf{i} - 12\,\mathbf{j} + 36\,\mathbf{k}) + (32\,\mathbf{i} - 16\,\mathbf{j} + 32\,\mathbf{k})$$

$$= 50\,\mathbf{i} - 28\,\mathbf{j} + 68\,\mathbf{k}\text{ N.}$$

Solve for the acceleration vector **a**, where

$$\mathbf{a} = \frac{\mathbf{F}_{\text{net}}}{m} = \frac{1}{8}(50\,\mathbf{i} - 28\,\mathbf{j} + 68\,\mathbf{k}) = 6.25\,\mathbf{i} - 3.5\,\mathbf{j} + 8.5\,\mathbf{k}\text{ m/s}^2.$$

Therefore the magnitude of the acceleration is given by

$$|\,\mathbf{a}\,| = \sqrt{6.25^2 + (-3.5)^2 + 8.5^2} = 11.116 \text{ m/s}^2.$$

Vector Algebra Tools Solution

Solving for the magnitude of the acceleration vector, $|\,\mathbf{a}\,|$.

Using the Vector Composition Tool: Enter the given value for the magnitude of $\mathbf{F}_1$, $|\,\mathbf{F}_1\,|$, in the Magnitude of vector **a**, $|\,\mathbf{a}\,|$ user input field. Enter the given components of vector $\mathbf{d}_1$ that is in the direction of $\mathbf{F}_1$ in the Vector components of vector **n** that is in the direction of vector **a** user input fields. Then click the blue Vector **a** button. The answers will be displayed in the output fields as shown. Click the Reset Values button before continuing to the next step.

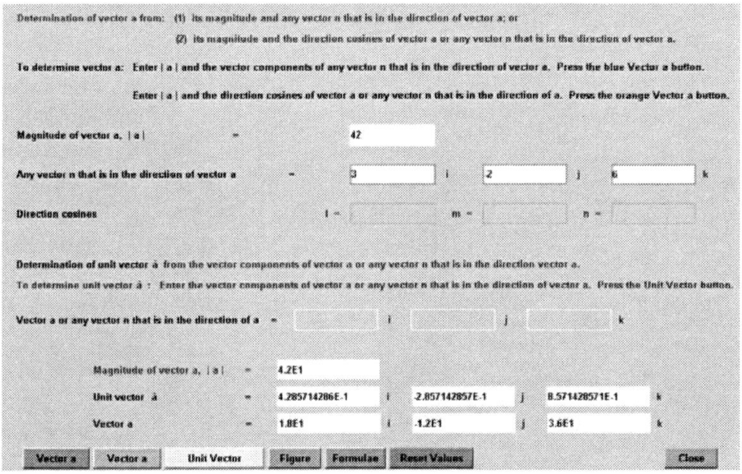

Solution: $\mathbf{F}_1 = 18\,\mathbf{i} - 12\,\mathbf{j} + 36\,\mathbf{k}$ N.

Using the Vector Composition Tool: Enter the given value for the magnitude of $\mathbf{F}_2$, $|\,\mathbf{F}_2\,|$, in the Magnitude of vector **a**, $|\,\mathbf{a}\,|$ user input field. Enter the given components of vector $\mathbf{d}_2$ that is in the direction of $\mathbf{F}_2$ in the Vector components of vector **n** that is in the direction of vector **a** user input fields. Then click the blue Vector **a** button. The answers will be displayed in the output fields as shown.

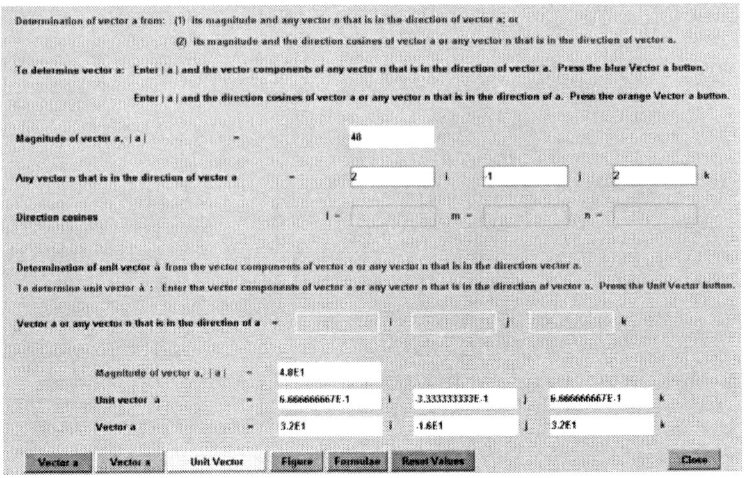

Determination of vector a from: (1) its magnitude and any vector n that is in the direction of vector a; or
(2) its magnitude and the direction cosines of vector a or any vector n that is in the direction of vector a.

To determine vector a: Enter | a | and the vector components of any vector n that is in the direction of vector a. Press the blue Vector a button.
Enter | a | and the direction cosines of vector a or any vector n that is in the direction of a. Press the orange Vector a button.

Magnitude of vector a, | a | = 48

Any vector n that is in the direction of vector a = 2 i -1 j 2 k

Direction cosines l = m = n =

Determination of unit vector â from the vector components of vector a or any vector n that is in the direction vector a.
To determine unit vector â : Enter the vector components of vector a or any vector n that is in the direction of vector a. Press the Unit Vector button.

Vector a or any vector n that is in the direction of a = i j k

Magnitude of vector a, | a | = 4.8E1
Unit vector â = 6.6666666667E-1 i -3.3333333333E-1 j 6.6666666667E-1 k
Vector a = 3.2E1 i -1.6E1 j 3.2E1 k

Vector a | Vector a | Unit Vector | Figure | Formulae | Reset Values | Close

Solution: $\mathbf{F}_2 = 32\,\mathbf{i} - 16\,\mathbf{j} + 32\,\mathbf{k}$ N.

Using the Vector Arithmetic Tool: Enter the calculated components of vector $\mathbf{F}_1$ in the Vector **a** user input fields, and enter the calculated components of vector $\mathbf{F}_2$ in the Vector **b** user input fields. Then enter the scalar multiplier, which is reciprocal value of the given value of the mass, in both the Scalar multiplier, p and q user input fields. Then click the Add **a** + **b** button. The answers will be displayed in the output fields as shown.

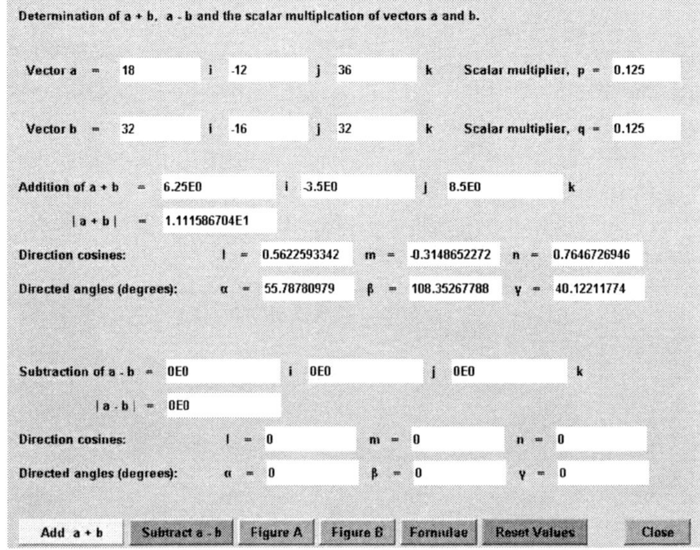

Solution: 4(a) Acceleration of the object, $|\,\mathbf{a}\,| = 11.116$ m/s^2.

4(b) **Worked Solution**

The work done by $\mathbf{F}_{\text{net}}$ that moves the object along the displacement vector **d** is given by the expression $W = \mathbf{F}_{\text{net}} \bullet \mathbf{d}$.

Solve for the displacement vector **d**, where

$$\mathbf{d} = \mathbf{p} - \mathbf{s} = (11\,\mathbf{i} - 7\,\mathbf{j} - 6\,\mathbf{k}) - (2\,\mathbf{i} + 3\,\mathbf{j} - 4\,\mathbf{k}) = 9\,\mathbf{i} - 10\,\mathbf{j} - 2\,\mathbf{k}\text{ m.}$$

Hence, the work W done by $\mathbf{F}_{\text{net}}$ is given by

$$\begin{aligned} W &= \mathbf{F}_{\text{net}} \bullet \mathbf{d} \\ &= (50\,\mathbf{i} - 28\,\mathbf{j} + 68\,\mathbf{k}) \bullet (9\,\mathbf{i} - 10\,\mathbf{j} - 2\,\mathbf{k}) \\ &= 450 + 280 - 136 \\ &= 594\text{ J.} \end{aligned}$$

Vector Algebra Tools Solution

Solving for the work done by $\mathbf{F}_{net}$.

Using the Vector Arithmetic Tool: Enter the given components of vector **p** in the Vector **a** user input fields and the given components of vector **s** in the Vector **b** user input fields. Then click the Subtract **a** – **b** button to determine the displacement vector **d**. The answers will be displayed in the output fields as shown. Click the Reset Values button before continuing to the next step.

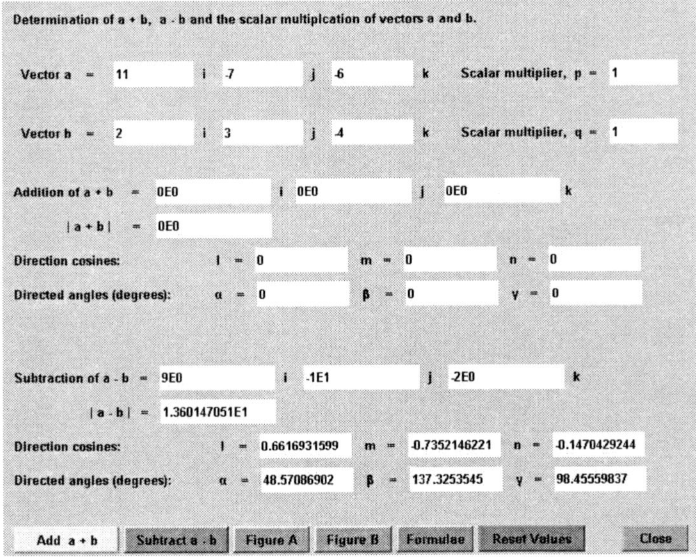

Solution: Displacement, $\mathbf{d} = \mathbf{s} - \mathbf{p} = 9\,\mathbf{i} - 10\,\mathbf{j} - 2\,\mathbf{k}$ m.

Using the Vector Arithmetic Tool: Enter the calculated components of vector $\mathbf{F}_1$ in the Vector **a** user input fields, and enter the calculated components of vector $\mathbf{F}_2$ in the Vector **b** user input fields. Then click the Add **a** + **b** button to determine $\mathbf{F}_{net}$. The answers will be displayed in the output fields as shown.

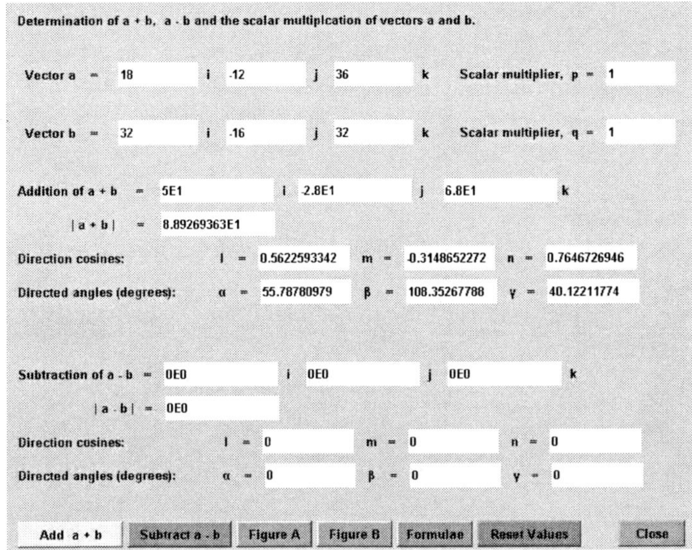

Solution: Force, $\mathbf{F}_{net} = \mathbf{F}_1 + \mathbf{F}_2 = 50\,\mathbf{i} - 28\,\mathbf{j} + 68\,\mathbf{k}$ N.

Using the Scalar Product Tool: Enter the calculated components of vector $\mathbf{F}_{net}$ in the Vector **a** user input fields. Enter the calculated components of vector **d** in the Vector **b** user input fields. Then click the Scalar Product **a** • **b** button to determine *W*. The answers will be displayed in the output fields as shown.

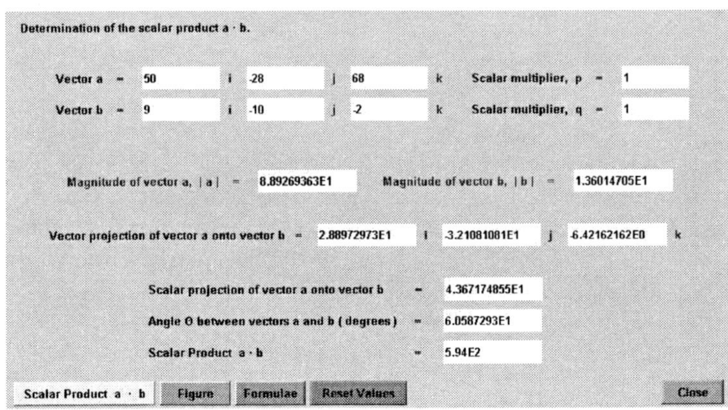

Solution: 4(b) Work, $W = 594$ J.

Answers: 4(a) Acceleration, $|\,\mathbf{a}\,| = 11.116$ m/s^2 and 4(b) Work, $W = 594$ J.

Question 5: Determine the angle between the pair of straight lines, whose vector equations are given by

$$\mathbf{r} = 2\,\mathbf{i} + 3\,\mathbf{j} - 2\,\mathbf{k} + \lambda(2\,\mathbf{i} - 5\,\mathbf{j} + 2\,\mathbf{k});$$
$$\mathbf{r} = 3\,\mathbf{i} - \mathbf{j} - \mathbf{k} + \mu(4\,\mathbf{i} - 3\,\mathbf{j} + 5\,\mathbf{k}).$$

Worked Solution

The vector equation of a straight line that passes through a fixed point with position vector **a** and parallel to vector **b** is given by $\mathbf{r} = \mathbf{a} + \lambda\mathbf{b}$. Hence, the angle θ between the two straight lines can be expressed as

$$\theta = \cos^{-1}\left(\frac{\mathbf{b}_1 \bullet \mathbf{b}_2}{|\mathbf{b}_1||\mathbf{b}_2|}\right),$$

where $\mathbf{b}_1$ and $\mathbf{b}_2$ are the direction vectors of each line.
Therefore, given $\mathbf{b}_1 = 2\,\mathbf{i} - 5\,\mathbf{j} + 2\,\mathbf{k}$ and $\mathbf{b}_2 = 4\,\mathbf{i} - 3\,\mathbf{j} + 5\,\mathbf{k}$, then the angle θ between the two straight lines is

$$\theta = \cos^{-1}\left(\frac{(2\,\mathbf{i} - 5\,\mathbf{j} + 2\,\mathbf{k})\,(4\,\mathbf{i} - 3\,\mathbf{j} + 5\,\mathbf{k})}{\sqrt{2^2 + (-5)^2 + 2^2}\ \sqrt{4^2 + (-3)^2 + 5^2}}\right)$$

$$= \cos^{-1}\left(\frac{33}{\sqrt{2^2 + (-5)^2 + 2^2}\ \sqrt{4^2 + (-3)^2 + 5^2}}\right)$$

$$= \cos^{-1}\left(\frac{33}{40.62}\right) = 35.67°.$$

Vector Algebra Tools Solution

Using the Scalar Product Tool: Enter the given components of vector $\mathbf{b}_1$ in the Vector **a** user input fields and the given components of vector $\mathbf{b}_2$ in the Vector **b** user input fields. Then click the Scalar Product $\mathbf{a} \bullet \mathbf{b}$ button. The answers will be displayed in the output fields as shown.

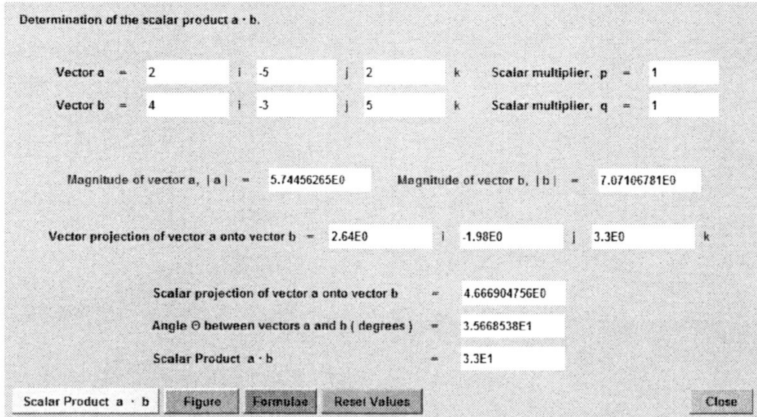

Answer: The angle between the two straight lines is $\theta = 35.67°$.

Question 6: Determine the angle between the straight line $\mathbf{r} = 2\,\mathbf{i} + 3\,\mathbf{j} - 2\,\mathbf{k} + \lambda(2\,\mathbf{i} - 5\,\mathrm{j} + 2\,\mathbf{k})$ and the vector normal to plane $\mathbf{r} \bullet (-2\,\mathbf{i} - 6\,\mathbf{j} - 3\,\mathbf{k}) = 1$.

Worked Solution

The vector equation of a straight line that passes through a fixed point with position vector **a** and is parallel to vector **b** is given by $\mathbf{r} = \mathbf{a} + \lambda\mathbf{b}$. The vector form of the equation of the plane in scalar form is expressed as $\mathbf{r} \bullet \mathbf{n} = d$, where **n** is the vector normal to plane. Referring to figure 4.10, the angle ϕ between the line and the vector normal to the plane is given by

$$\cos\phi = \left(\frac{\mathbf{b} \bullet \mathbf{n}}{|\mathbf{b}||\mathbf{n}|}\right).$$

Since **n** is perpendicular to the plane, the angle θ between the line and the plane is given by $90° - \phi$. Alternatively, the angle θ may be calculated using $\sin\theta = \cos\phi$.

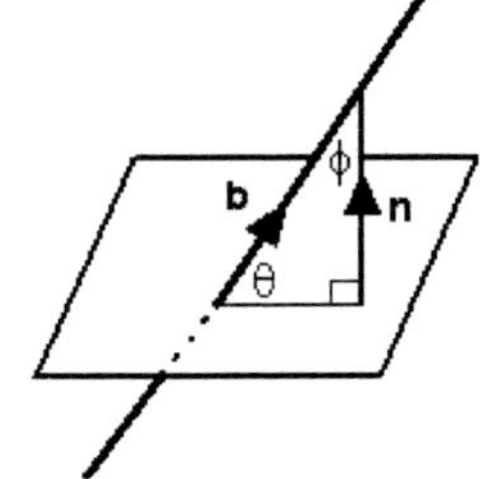

Fig. 4.10

From the vector equations of the line and the plane, we can write $\mathbf{b} = 2\,\mathbf{i} - 5\,\mathrm{j} + 2\,\mathbf{k}$ and $\mathbf{n} = -2\,\mathbf{i} - 6\,\mathbf{j} - 3\,\mathbf{k}$.

Hence, the angle ϕ between the straight line and the plane is

$$\cos\phi = \left(\frac{\mathbf{b} \bullet \mathbf{n}}{|\mathbf{b}||\mathbf{n}|}\right) = \left(\frac{(2\,\mathbf{i} - 5\,\mathbf{j} + 2\,\mathbf{k})(-2\,\mathbf{i} - 6\,\mathbf{j}\, - 3\,\mathbf{k})}{\sqrt{2^2 + (-5)^2 + 2^2}\,\sqrt{(-2)^2 + (-6)^2 + (-3)^2}}\right)$$

$$= \left(\frac{20}{\sqrt{33}\,\sqrt{49}}\right) = \left(\frac{20}{7\sqrt{33}}\right).$$

Therefore,

$$\phi = \cos^{-1}\left(\frac{20}{7\sqrt{33}}\right) = 60.17°.$$

Note that the angle between the line and the plane is $\theta = 90° - 60.17° = 29.83°$.

Vector Algebra Tools Solution

Using the Scalar Product Tool: Enter the given components of vector **b** in the Vector **a** user input fields and the given components of vector **n** in the Vector **b** user input fields. Then click the Scalar Product **a • b** button. The answers will be displayed in the output fields as shown.

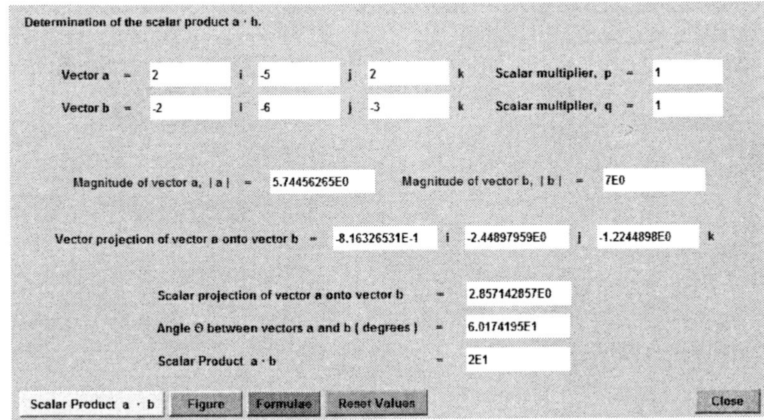

Answer: The angle between straight line and vector normal to the plane is $\phi = 60.17°$.

Question 7: The uniform electric field in the region of a plane surface is given by $\mathbf{E} = 3\times10^3\,(2\,\mathbf{i} - \mathbf{j} + 3\,\mathbf{k})$ N/C. If the directed area vector normal to the surface of the plane is $\mathbf{A} = 0.25\,\mathbf{i} + 0.15\,\mathbf{j} + 0.25\,\mathbf{k}$ m^2, calculate the electric flux through the surface.

Worked Solution

The electric flux Φ_E through the surface is given by

$$\Phi_E = \mathbf{E} \bullet \mathbf{A},$$

where $\mathbf{E} = 3\times10^3\,(2\,\mathbf{i} - \mathbf{j} + 3\,\mathbf{k})$ N/C and $\mathbf{A} = 0.25\,\mathbf{i} + 0.15\,\mathbf{j} + 0.25\,\mathbf{k}$ m^2.

Therefore,

$$\Phi_E = 3\times10^3\,(2\,\mathbf{i} - \mathbf{j} + 3\,\mathbf{k}) \bullet (0.25\,\mathbf{i} + 0.15\,\mathbf{j} + 0.25\,\mathbf{k})$$

$$= 1500 - 450 + 2250 = 3.3\times10^3 \text{ N m}^2/\text{C}.$$

Vector Algebra Tools Solution

Using the Scalar Product Tool: Enter the components of vector **E** in the Vector **a** user input fields and the components of vector **A** in the Vector **b** user input fields. Then click the Scalar Product **a** • **b** button to determine Φ_E. The answers will be displayed in the output fields as shown.

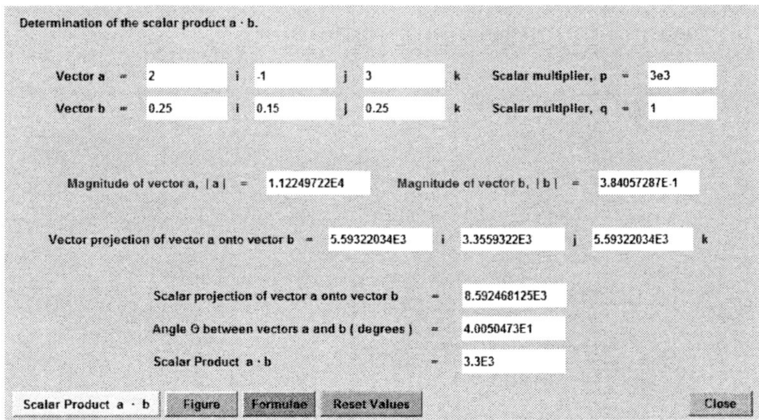

Answer: The electric flux through the surface is $\Phi_E = 3.3\times10^3$ N m^2/C.

4.7 NOTES:

CHAPTER 5

VECTOR PRODUCT

In this chapter the vector product, which is the product of two vectors resulting in a vector quantity, is presented. Specifically, vector product definitions, properties of vector product algebra, problem solving applications and worked examples are included. Also demonstrated is the use of the Vector Product and the Forces on an Electric Charge software tools in solving the example problems.

5.1 DEFINITION OF THE VECTOR PRODUCT

The **vector product** or **cross product** of any two vectors **a** and **b**, whose directions are at an angle θ to each other, as shown in figure 5.1, is given as the vector **c**, where

$$\mathbf{c} = \mathbf{a} \times \mathbf{b} = |\,\mathbf{a}\,|\,|\,\mathbf{b}\,|\sin\theta\ \hat{\mathbf{n}},\ \text{where } 0° \le \theta \le 180°.$$

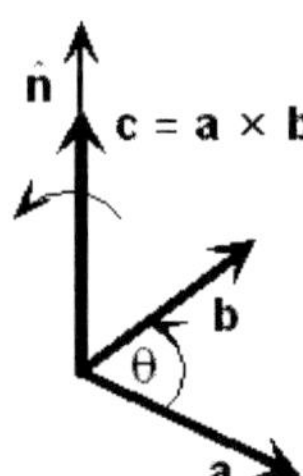

Fig. 5.1. The direction of the vector product **c** is normal to the plane containing vectors **a** and **b** and is governed by the right-hand rule.

The vector product **c** is the product of the magnitude, $|\,\mathbf{a}\,|\,|\,\mathbf{b}\,|\sin\theta$, and the unit vector $\hat{\mathbf{n}}$, whose direction is perpendicular to the directions of both vector **a** and vector **b**. The direction of the vector product **c** is determined using the **right-hand rule**, where the vectors **a**, **b** and $\hat{\mathbf{n}}$, taken in this order, form a right-handed coordinate system. To apply the right-hand rule, first extend your right hand and point your fingers to align with vector **a**. Next, without moving your thumb, curl your fingers inwards towards vector **b** in the direction of the angle θ. Your thumb then points in the direction of vector **c**, which is perpendicular or normal to the plane containing the vectors **a** and **b**.

The magnitude of the vector product is given by the expression

$$|\,\mathbf{c}\,| = |\,\mathbf{a} \times \mathbf{b}\,| = |\,\mathbf{a}\,|\,|\,\mathbf{b}\,|\sin\theta = ab\sin\theta,\ \text{where } 0° \le \theta \le 180°.$$

Geometrically, the magnitude of the vector product $\mathbf{a} \times \mathbf{b}$ is represented by the area of a parallelogram having vectors **a** and **b** as its sides. See figure 5.8.

5.2 DETERMINATION OF VECTOR PRODUCT QUANTITIES

5.2.1 Vector product

If vectors **a** and **b** are represented in their component form in a three-dimensional Cartesian coordinate system, where $\mathbf{a} = a_x\,\mathbf{i} + a_y\,\mathbf{j} + a_z\,\mathbf{k}$ and $\mathbf{b} = b_x\,\mathbf{i} + b_y\,\mathbf{j} + b_z\,\mathbf{k}$, then the vector product of the vectors **a** and **b** can be determined algebraically as follows:

$$\mathbf{c} = \mathbf{a} \times \mathbf{b} = (a_y b_z - a_z b_y)\,\mathbf{i} + (a_z b_x - a_x b_z)\,\mathbf{j} + (a_x b_y - a_y b_x)\,\mathbf{k}.$$

This equation can be conveniently expressed in determinant form, where the components of vectors **a** and **b** are written as the second and third rows of a 3 × 3 determinant as follows:

$$\mathbf{c} = \mathbf{a} \times \mathbf{b} = \begin{vmatrix} \mathbf{i} & \mathbf{j} & \mathbf{k} \\ a_x & a_y & a_z \\ b_x & b_y & b_z \end{vmatrix}.$$

Based upon the properties of determinants, the interchange of rows two and three of the determinant corresponding to the vector product $\mathbf{b} \times \mathbf{a}$, changes the sign of the determinant, such that $\mathbf{a} \times \mathbf{b} = -\mathbf{b} \times \mathbf{a}$.

5.2.2 Magnitude of the vector product

The magnitude of the vector product of vectors **a** and **b** is expressed algebraically as

$$|\,\mathbf{a} \times \mathbf{b}\,| = \sqrt{(a_y b_z - a_z b_y)^2 + (a_z b_x - a_x b_z)^2 + (a_x b_y - a_y b_x)^2}\,.$$

5.2.3 The angle between two vectors

The angle θ between the directions of vectors **a** and **b** is given by

$$\theta = \sin^{-1}\left(\frac{|\mathbf{a} \times \mathbf{b}|}{|\mathbf{a}|\,|\mathbf{b}|}\right)$$

$$= \sin^{-1}\left(\sqrt{\frac{(a_yb_z - a_zb_y)^2 + (a_zb_x - a_xb_z)^2 + (a_xb_y - a_yb_x)^2}{(a_x^2 + a_y^2 + a_z^2)(b_x^2 + b_y^2 + b_z^2)}}\right),$$

where $0° \leq \theta \leq 180°$.

5.3 LAWS AND PROPERTIES OF VECTOR PRODUCT ALGEBRA

Important laws and properties in vector product algebra for the nonzero vectors **a**, **b** and **c** and the nonzero real number scalar quantities p and q are summarized below.

$\mathbf{a} \times \mathbf{b} \neq \mathbf{b} \times \mathbf{a}$	Commutative law does *not* apply
$\mathbf{a} \times \mathbf{b} = -\mathbf{b} \times \mathbf{a} = \mathbf{b} \times -\mathbf{a} = -(\mathbf{b} \times \mathbf{a})$	
$\mathbf{a} \times (\mathbf{b} \times \mathbf{c}) \neq (\mathbf{a} \times \mathbf{b}) \times \mathbf{c}$	Associative law does *not* apply
$\mathbf{a} \times (\mathbf{b} + \mathbf{c}) = (\mathbf{a} \times \mathbf{b}) + (\mathbf{a} \times \mathbf{c})$	Distributive law
$p\mathbf{a} \times q\mathbf{b} = pq(\mathbf{a} \times \mathbf{b})$	
$p(\mathbf{a} \times \mathbf{b}) = (p\mathbf{a}) \times \mathbf{b} = \mathbf{a} \times (p\mathbf{b}) = (\mathbf{a} \times \mathbf{b})p$	
$(p\mathbf{a} + q\mathbf{b}) \times \mathbf{c} = p(\mathbf{a} \times \mathbf{c}) + q(\mathbf{b} \times \mathbf{c})$	

The right-hand rule governs the direction of the vector product and clearly illustrates its noncommutative property. For example, when the order in which the vectors **a** and **b** appear in the vector product are reversed, the sign of the vector product is changed and, hence, its direction. The vectors $\mathbf{a} \times (\mathbf{b} \times \mathbf{c})$ and $(\mathbf{a} \times \mathbf{b}) \times \mathbf{c}$ are the vector triple products of three vectors and are discussed in Chapter 6.

5.4 IMPORTANT RESULTS INVOLVING VECTOR PRODUCTS

Important results and operations involving vector products are outlined as follows:

- If either one or both of the vectors **a** and **b** are zero vectors, then $\mathbf{a} \times \mathbf{b} = \mathbf{0}$. However, if the nonzero vectors **a** and **b** are **parallel**, pointing in same direction, then the angle θ between the directions of the two vectors is given by $\theta = 0°$, such that $\sin 0° = 0$ and $\mathbf{a} \times \mathbf{b} = \mathbf{0}$. Conversely, if the nonzero vectors **a** and **b** are **antiparallel**, pointing in opposite directions, then the angle θ between the directions of the two vectors is given by $\theta = 180°$, such that $\sin 180° = 0$ and $\mathbf{a} \times \mathbf{b} = \mathbf{0}$. Therefore, if $\mathbf{a} \times \mathbf{b} = \mathbf{0}$, then **a** and **b** are **collinear**, linearly dependent vectors.

In the special case when both vectors in the vector product are identical, then the vector product is zero, such that

$$\mathbf{a} \times \mathbf{a} = \mathbf{b} \times \mathbf{b} = \mathbf{0}.$$

Note that vectors **a** and **b** are linearly independent if, and only if, $\mathbf{a} \times \mathbf{b} \neq \mathbf{0}$

- If the nonzero vectors **a** and **b** are orthogonal, i.e., perpendicular to each other, then the angle θ between the two vectors is $\theta = 90°$, such that $\sin 90° = 1$ and $|\,\mathbf{a} \times \mathbf{b}\,| = |\,\mathbf{a}\,|\,|\,\mathbf{b}\,| = ab$. This quantity is the scalar area of a rectangular with vectors **a** and **b** as its sides.

- If vectors **a** and **b** are noncollinear vectors that lie in the same plane, then a normal or orthogonal vector **n**, perpendicular to the plane, is given by

$$\mathbf{n} = \mathbf{a} \times \mathbf{b}.$$

Since vectors **a** and **b** are coplanar, then **n** is also orthogonal to vector **a** and vector **b**. Thus, to test that the nonzero vectors **a**, **b** and **n** are orthogonal, we use the **orthogonal condition**

$$\mathbf{a} \bullet \mathbf{n} = 0 \text{ and } \mathbf{b} \bullet \mathbf{n} = 0.$$

The orthogonal condition also provides a convenient test for checking the solutions in vector product calculations. See question 1 of the Worked Examples in section 5.6.

- If **i**, **j** and **k** are Cartesian unit vectors, then the vector products of two unit vectors, which are parallel, pointing in the same direction, are given by

$$\mathbf{i} \times \mathbf{i} = \mathbf{j} \times \mathbf{j} = \mathbf{k} \times \mathbf{k} = \mathbf{0}.$$

Likewise, the vector products of two unit vectors that are antiparallel, pointing in opposite directions, are given by

$$-\mathbf{i} \times \mathbf{i} = -\mathbf{j} \times \mathbf{j} = -\mathbf{k} \times \mathbf{k} = \mathbf{0} \text{ and } \mathbf{i} \times -\mathbf{i} = \mathbf{j} \times -\mathbf{j} = \mathbf{k} \times -\mathbf{k} = \mathbf{0}.$$

In contrast, the vector products of any two mutually perpendicular unit vectors are given by

$$\mathbf{j} \times \mathbf{k} = \mathbf{i}, \quad \mathbf{k} \times \mathbf{i} = \mathbf{j}, \quad \mathbf{i} \times \mathbf{j} = \mathbf{k},$$

$$\mathbf{k} \times \mathbf{j} = -\mathbf{i}, \quad \mathbf{i} \times \mathbf{k} = -\mathbf{j} \text{ and } \mathbf{j} \times \mathbf{i} = -\mathbf{k}.$$

Clearly, since the vector product operation is not commutative, changing the order in which the unit vectors are calculated changes the sign of the vector product and hence, its direction.

- Although the vector product of the vectors **a** and **b** is not governed by the commutative law of vector algebra, the magnitude of **a** × **b** is equal to the magnitude of **b** × **a**, and can be written as

$$|\mathbf{a} \times \mathbf{b}| = |\mathbf{b} \times \mathbf{a}|.$$

- If **a** and **b** are nonzero vectors then the Lagrange identity is given by

$$(\mathbf{a} \times \mathbf{b}) \bullet (\mathbf{a} \times \mathbf{b}) = |\mathbf{a} \times \mathbf{b}|^2 = a^2b^2 - (\mathbf{a} \bullet \mathbf{b})^2.$$

- If vectors **a**, **b**, **c** and **d** are nonzero vectors, then the generalized form of the Lagrange identity is given as

$$(\mathbf{a} \times \mathbf{b}) \bullet (\mathbf{c} \times \mathbf{d}) = (\mathbf{a} \bullet \mathbf{c})(\mathbf{b} \bullet \mathbf{d}) - (\mathbf{a} \bullet \mathbf{d})(\mathbf{b} \bullet \mathbf{c}).$$

This expression can also be written in the form of a determinant and is given by

$$\begin{vmatrix} \mathbf{a} \bullet \mathbf{c} & \mathbf{a} \bullet \mathbf{d} \\ \mathbf{b} \bullet \mathbf{c} & \mathbf{b} \bullet \mathbf{d} \end{vmatrix},$$

which reduces to the Lagrange identity when **c** = **a** and **d** = **b**.

- If **a** and **b** are nonzero vectors, then

$$\begin{aligned}(\mathbf{a} - \mathbf{b}) \times (\mathbf{a} + \mathbf{b}) &= \mathbf{a} \times \mathbf{a} + (-\mathbf{b} \times \mathbf{a}) + (\mathbf{a} \times \mathbf{b}) - \mathbf{b} \times \mathbf{b} \\ &= \mathbf{a} \times \mathbf{b} + \mathbf{a} \times \mathbf{b} \\ &= 2(\mathbf{a} \times \mathbf{b}) = 2\mathbf{a} \times \mathbf{b} = \mathbf{a} \times 2\mathbf{b}.\end{aligned}$$

5.5 PROBLEM SOLVING APPLICATIONS OF VECTOR PRODUCTS

The following examples demonstrate the numerous problem solving applications of vector products:

- The vector moment of the force **F** acting on a particle that is rotating about a fixed axis in an inertial frame of reference is given by the torque $\boldsymbol{\tau} = \mathbf{r} \times \mathbf{F}$, where **r** is the position vector of the particle P measured from the axis of rotation, i.e., the pivot point O, as shown figure 5.2.

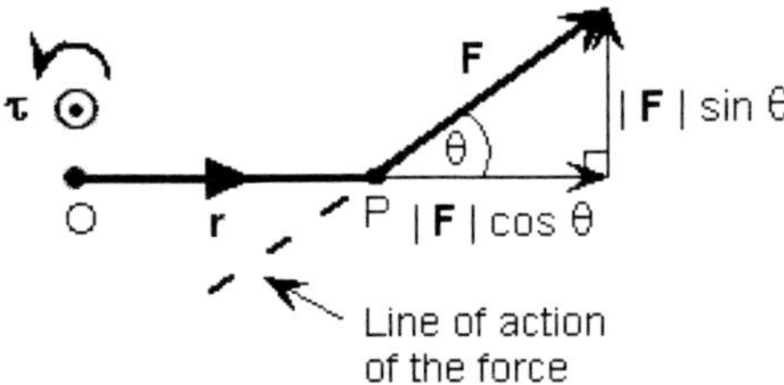

Fig. 5.2. The vector moment of the force **F** is given by the vector product $\mathbf{r} \times \mathbf{F}$. The quantity $|\mathbf{F}| \sin\theta$ is the moment arm of the force **F** and is the perpendicular distance from the rotational axis to the line of action of **F**.

The direction of the torque $\boldsymbol{\tau}$ is determined by the right-hand rule and is normal to the plane containing the vectors **r** and **F**. In particular, the rotation of P about O is counterclockwise and the sign of the torque, by convention, is considered positive. The magnitude of the torque is given by $|\boldsymbol{\tau}| = |\mathbf{r}|(|\mathbf{F}| \sin\theta)$, where θ is the angle between **r** and **F**. Specifically, the torque is at its maximum and is given by rF when $\theta = 90°$, and is zero when $\theta = 0°$ or $180°$. In addition, if the net external torque $\boldsymbol{\tau}_{net}$ about any point is zero, then the particle is in **rotational equilibrium**, such that $\boldsymbol{\tau}_{net} = \sum \boldsymbol{\tau} = \mathbf{0}$ and the angular momentum of the system is conserved.

- The vector moment of the resultant force $\mathbf{F}_{net}$ about a point O, i.e., the net external torque $\boldsymbol{\tau}_{net}$, is given by the vector sum of the individual vector moments of the forces, $\mathbf{F}_1$, $\mathbf{F}_2$, $\mathbf{F}_3$, . . . , $\mathbf{F}_n$, acting through a point P and can be expressed as

$$\boldsymbol{\tau}_{net} = \mathbf{r} \times \mathbf{F}_{net} = \sum \boldsymbol{\tau} = \mathbf{r} \times \mathbf{F}_1 + \mathbf{r} \times \mathbf{F}_2 + \mathbf{r} \times \mathbf{F}_3 + \cdots + \mathbf{r} \times \mathbf{F}_n$$

$$= \mathbf{r} \times (\mathbf{F}_1 + \mathbf{F}_2, + \mathbf{F}_3 + \cdots + \mathbf{F}_n),$$

where $\mathbf{r}$ is the position vector of the point P on the line of action of the forces with respect to O.

- The instantaneous linear velocity $\mathbf{v}$ of a particle that is rotating about a fixed axis in an inertial frame of reference is given by $\mathbf{v} = \boldsymbol{\omega} \times \mathbf{r}$, where $\boldsymbol{\omega}$ is the vector angular velocity of the particle and $\mathbf{r}$ is its position vector measured from the axis of rotation. Note that the direction of $\boldsymbol{\omega}$ is determined by the right-hand rule, such that when the fingers of your right hand are curled in the direction of the rotation of the particle, your extended thumb points in the direction of $\boldsymbol{\omega}$ that is along the axis of rotation. By convention, a counterclockwise rotation indicates a positive vector angular velocity, and a clockwise rotation indicates a negative vector angular velocity.

- The instantaneous angular momentum $\mathbf{L}$ of a particle that is rotating about a fixed axis in an inertial frame of reference is given by $\mathbf{L} = \mathbf{r} \times \mathbf{p}$, where $\mathbf{r}$ is the position vector of the particle measured from the axis of rotation and $\mathbf{p} = m\mathbf{v}$ is its linear momentum, where m is the mass and $\mathbf{v}$ is the instantaneous velocity of the particle. The magnitude of the angular momentum is given by $|\,\mathbf{L}\,| = |\,\mathbf{r} \times \mathbf{p}\,| = m\,|\,\mathbf{r}\,|\,(|\,\mathbf{v}\,|\sin\theta)$, where θ is the angle between $\mathbf{r}$ and $\mathbf{v}$. Specifically, the angular momentum is at its maximum and is given by mrv when $\theta = 90°$, and is zero when $\theta = 0°$ or $180°$. Also, the rate of change of the angular momentum of a particle about the axis of rotation is equal to the vector moment of the net force $\mathbf{F}_{\text{net}}$ acting on the particle and is given by $\dfrac{d\mathbf{L}}{dt} = \mathbf{r} \times \mathbf{F}_{\text{net}} = \boldsymbol{\tau}_{\text{net}}$. Furthermore, if the moment of the net force about the axis of rotation is zero, i.e., there is no net external torque acting on the system, then $\mathbf{L}$ is a constant and the angular momentum of the system is conserved.

- The magnetic force on a moving electric charge in a uniform external magnetic field is given by $\mathbf{F}_B = q(\mathbf{v} \times \mathbf{B})$, where q is the electric charge, $\mathbf{v}$ is the instantaneous velocity of the charge and $\mathbf{B}$ is the magnetic field. The direction of $\mathbf{F}_B$ exerted on a positively charged particle is in the same direction as $\mathbf{v} \times \mathbf{B}$ and is in the opposite direction of $\mathbf{v} \times \mathbf{B}$ for a negatively charged particle. In both cases $\mathbf{F}_B$ is perpendicular to the plane containing the vectors $\mathbf{v}$ and $\mathbf{B}$. See figure 5.3. The magnitude of $\mathbf{F}_B$ is given by $|\,\mathbf{F}_B\,| = |\,q\,|\,|\,\mathbf{B}\,|\,(|\,\mathbf{v}\,|\sin\theta)$, where θ is the angle between $\mathbf{v}$ and $\mathbf{B}$.

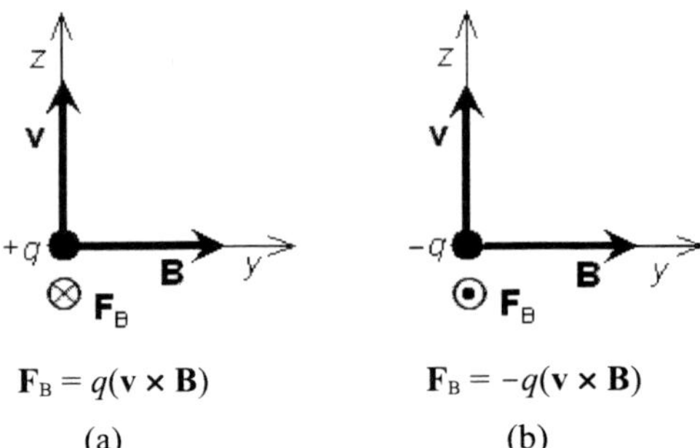

$\mathbf{F}_B = q(\mathbf{v} \times \mathbf{B})$

(a)

$\mathbf{F}_B = -q(\mathbf{v} \times \mathbf{B})$

(b)

Fig. 5.3. The magnetic force $\mathbf{F}_B$ on (a) a positively charged particle and (b) a negatively charged particle, moving with a constant velocity **v** in a uniform external magnetic field **B**.

- The **Lorentz force**, which is the net force acting on a moving charge q in an external uniform electric and magnetic field is given by the expression $\mathbf{F}_L = q\mathbf{E} + q(\mathbf{v} \times \mathbf{B})$, where q is the electric charge, **v** is the instantaneous velocity of the charge and **E** and **B** are the electric and magnetic fields, respectively.

- The magnetic force on a straight line conductor in a uniform external magnetic field **B**, as shown in figure 5.4, is given by $\mathbf{F}_B = I(\mathbf{l} \times \mathbf{B})$, where I is the current in the conductor and **l** is the vector line element of the conductor that points in the direction of conventional current flow, i.e., the positive charge, and whose magnitude is the length of the conductor. The magnitude of $\mathbf{F}_B$ is given by $|\mathbf{F}_B| = I|\mathbf{B}|(|\mathbf{l}|\sin\theta)$, where θ is the angle between **l** and **B**. Specifically, the maximum magnetic force occurs when $\theta = 90°$, such that **l** is perpendicular to **B**. In contrast, the magnetic force is zero when $\theta = 0°$ or $180°$, such that **l** is parallel or antiparallel to **B**. The direction of $\mathbf{F}_B$ is governed by the right-hand rule and is determined by pointing the fingers of your right-hand in the direction of the current I and then curling them towards **B**. Your upright thumb indicates the direction of $\mathbf{F}_B$ and is normal to the plane containing the vectors **l** and **B**.

Fig. 5.4. The magnetic force $\mathbf{F}_B$ on the straight line conductor is perpendicular to the x-y plane containing $\mathbf{l}$ and $\mathbf{B}$.

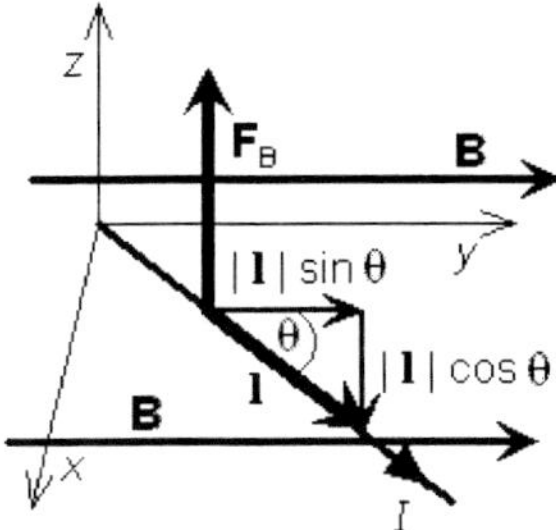

- Shown in plan view in figure 5.5 is the magnetic dipole of a rectangular current loop in a uniform external magnetic field **B**.

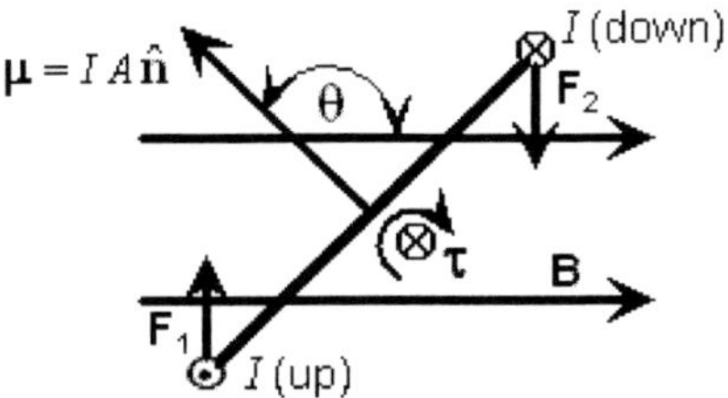

Fig. 5.5. The torque $\boldsymbol{\tau}$ on the magnetic dipole is directed into the page of the book and tends to align the coil so that $\boldsymbol{\mu}$ is parallel to $\mathbf{B}$, such that $\boldsymbol{\mu} \times \mathbf{B} = \mathbf{0}$.

The forces $\mathbf{F}_1$ and $\mathbf{F}_2$, acting on the two straight current carrying wire segments that are perpendicular to the **B**, are equal in magnitude but opposite in direction, such that the net force is zero. However, since each force has a different line of action, the pair of parallel forces constitutes a **couple** and results in a net torque on the magnetic dipole. For example, if the axis of rotation is about $\mathbf{F}_1$, then the torque $\boldsymbol{\tau}$ on the magnetic dipole due to the magnetic field is given by

$$\boldsymbol{\tau} = \boldsymbol{\mu} \times \mathbf{B}.$$

In this expression the magnetic dipole moment is defined by $\boldsymbol{\mu} = I\mathbf{A} = IA\,\hat{\mathbf{n}}$, where I is the current in the loop, A is the area of the loop and $\hat{\mathbf{n}}$ is the unit vector perpendicular to the plane of the loop. In addition, the direction of $\hat{\mathbf{n}}$ is governed by the right-hand rule and is determined by curling the fingers of your right hand in the direction of the current in the wire such that your upright thumb points in the direction of $\hat{\mathbf{n}}$. The magnitude of the torque is given by $|\boldsymbol{\tau}| = |\boldsymbol{\mu}|(|\mathbf{B}|\sin\theta)$, where θ is the angle between $\boldsymbol{\mu}$ and $\mathbf{B}$. Specifically, the torque is at its maximum when $\theta = 90°$, such that $\boldsymbol{\mu}$ is perpendicular to $\mathbf{B}$, and is zero when $\theta = 0°$ or $180°$, such that $\boldsymbol{\mu}$ is parallel or antiparallel to $\mathbf{B}$. Note that the torque $\boldsymbol{\tau}$ tends to align the magnetic dipole so that $\boldsymbol{\mu}$ is parallel to $\mathbf{B}$; hence, $\boldsymbol{\mu} \times \mathbf{B} = \mathbf{0}$.

- Shown in figure 5.6 is an electric dipole in a uniform external electric field **E**.

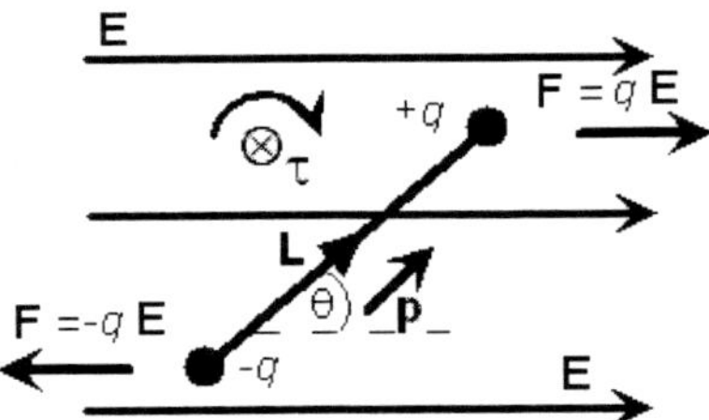

Fig. 5.6. The torque $\boldsymbol{\tau}$ on the electric dipole that is directed into the page of the book and tends to align the electric dipole so that $\mathbf{p}$ is parallel to $\mathbf{E}$, such that $\mathbf{p} \times \mathbf{E} = \mathbf{0}$.

The force on the positive charge of the electric dipole is $q\mathbf{E}$, and the force on the negative charge is $-q\mathbf{E}$. Both forces are equal in magnitude but opposite in direction. Hence, the net force on the electric dipole is zero, and the electric dipole is in translational equilibrium. However, since the forces have different lines of action, the pair of parallel forces constitutes a **couple** and results in a net torque on the electric dipole. For example, if the axis of rotation is about $-q$, then the torque $\boldsymbol{\tau}$ on the electric dipole due to the electric field is given by

$$\boldsymbol{\tau} = \mathbf{p} \times \mathbf{E},$$

where the electric dipole moment is given by $\mathbf{p} = |q|\mathbf{L}$, such that the absolute magnitude of the charge is $|q|$ and $\mathbf{L}$ is the displacement vector that points from $-q$ to $+q$. The direction of $\boldsymbol{\tau}$ is governed by the right-hand rule. Therefore, when the dipole is oriented as shown in figure 5.5, the torque, which points into the page of the book, rotates the electric dipole in a clockwise direction. The magnitude of the torque is given by $|\boldsymbol{\tau}| = |\mathbf{p}|(|\mathbf{E}|\sin\theta)$, where θ is the angle between **p** and **E**. Specifically, the torque is at its maximum when $\theta = 90°$ and **p** is perpendicular to **E**, and is zero when $\theta = 0°$ or $180°$and **p** is either parallel or antiparallel to **E**. Note that the torque $\boldsymbol{\tau}$ tends to align the magnetic dipole such that **p** is parallel with **E**; hence, $\mathbf{p} \times \mathbf{E} = \mathbf{0}$.

- The instantaneous magnetic field **B** produced by a moving electric charge q at a fixed point is given by $\mathbf{B} = (\mu_o q/4\pi r^3)(\mathbf{v} \times \mathbf{r})$, where μ_o is the permeability of free space and is given by $4\pi \times 10^{-7}$ H/m, **v** is the instantaneous velocity of the charge and **r** is the displacement vector of the fixed point from q.

- The rate of flow of energy, the energy flux, in a electromagnetic wave is along the direction of propagation and is defined by the Poynting vector **S**, which is given by the expression

$$\mathbf{S} = \frac{1}{\mu_o}(\mathbf{E} \times \mathbf{B}),$$

where μ_o is the permeability of free space and is given by $4\pi \times 10^{-7}$ H/m. Specifically, the **Poynting vector** represents the power per unit area, where the unit area is oriented at right angles to the direction of travel of the electromagnetic wave. See figure 5.7.

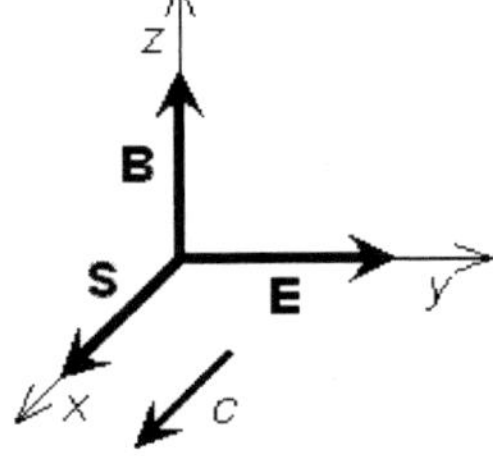

Fig. 5.7. The Poynting vector **S** for a plane electromagnetic wave in the direction of the x-axis is along the direction of propagation of the electromagnetic wave c.

- The area of a parallelogram, with vectors **a** and **b** as its sides, as shown in figure 5.8, is given by

$$|\mathbf{b}|h = |\mathbf{a}||\mathbf{b}|\sin\theta = |\mathbf{a} \times \mathbf{b}|,$$

where $|\mathbf{b}|$ is the length of the base and h is the altitude of the parallelogram. The vector area of the parallelogram is given by $\mathbf{a} \times \mathbf{b}$ and is normal to the plane of the parallelogram whose direction is determined by the right-hand rule.

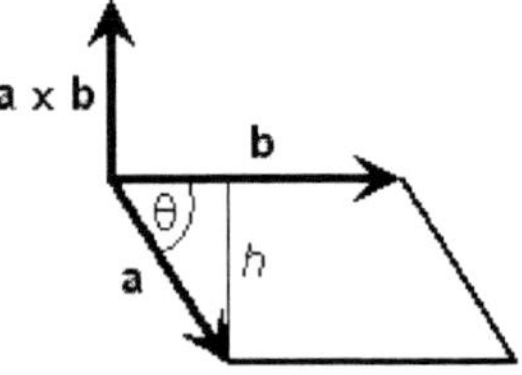

Fig. 5.8. The area of a parallelogram is given by $|\mathbf{a} \times \mathbf{b}|$.

- If vectors **a**, **b** and **c** represent the sides of the plane triangle ABC, as shown in figure 5.9, then the area of the triangle, Δ, is given by $\Delta = \frac{1}{2}|\mathbf{a} \times \mathbf{b}| = \frac{1}{2}|\mathbf{b} \times \mathbf{c}| = \frac{1}{2}|\mathbf{c} \times \mathbf{a}|$.

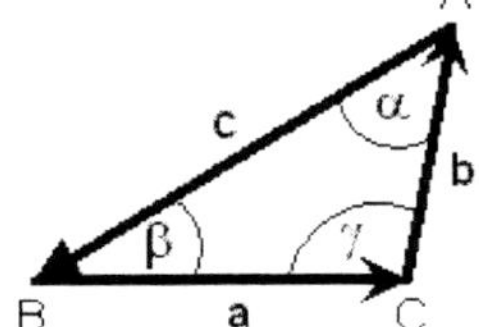

Fig. 5.9. The plane triangle ABC formed by the vectors **a**, **b** and **c**.

The vector area of the triangle is determined by the vector $\mathbf{\Delta} = \frac{1}{2}\mathbf{a} \times \mathbf{b} = \frac{1}{2}\mathbf{b} \times \mathbf{c} = \frac{1}{2}\mathbf{c} \times \mathbf{a}$, where $\mathbf{\Delta}$ is a vector normal to the plane of the triangle whose direction is determined by the right-hand rule.

- In general, if the vertices of a plane triangle ABC are given by the position vectors **a**, **b** and **c**, then the vector area of the triangle, $\mathbf{\Delta} = \frac{1}{2}(\mathbf{b} \times \mathbf{c} + \mathbf{c} \times \mathbf{a} + \mathbf{a} \times \mathbf{b})$. However, if the vertices of the triangle are collinear, then $\mathbf{\Delta} = \frac{1}{2}(\mathbf{b} \times \mathbf{c} + \mathbf{c} \times \mathbf{a} + \mathbf{a} \times \mathbf{b}) = \mathbf{0}$.

- For any plane triangle, the ratio of the length of a side and the sine of the angle opposite that side is a constant and is governed by the **law of sines**. This relationship is expressed by a determination of the area of the plane triangle shown in figure 5.9 as follows:

$$\frac{1}{2}|\mathbf{b}||\mathbf{c}|\sin\alpha = \frac{1}{2}|\mathbf{c}||\mathbf{a}|\sin\beta = \frac{1}{2}|\mathbf{a}||\mathbf{b}|\sin\gamma,$$

which is more commonly written as

$$\frac{a}{\sin\alpha} = \frac{b}{\sin\beta} = \frac{c}{\sin\gamma},$$

where $|\mathbf{a}| = a$, $|\mathbf{b}| = b$ and $|\mathbf{c}| = c$.

- If vectors **a**, **b**, **c** and **d** represent the sides of a plane quadrilateral ABCD, such that $\mathbf{a} + \mathbf{b} + \mathbf{c} + \mathbf{d} = \mathbf{0}$, then the vector area of the quadrilateral is given by $\frac{1}{2}(\mathbf{b} + \mathbf{c}) \times (\mathbf{a} + \mathbf{b})$.

- The shortest distance, d, between two parallel lines, $\mathbf{r} = \mathbf{a}_1 + \lambda\mathbf{b}$ and $\mathbf{r} = \mathbf{a}_2 + \mu\mathbf{b}$, is given by

$$d = \frac{1}{|\mathbf{b}|} \, |(\mathbf{a}_2 - \mathbf{a}_1) \times \mathbf{b}|.$$

5.6 WORKED EXAMPLES

The **Vector Product Tool** can be used to determine the vector product of any two vectors in component form in a three-dimensional Cartesian coordinate system. In addition, the Vector Product Tool can also be used to calculate the magnitude of the vector product, a unit vector that is in the direction of the vector product and the angle between the two vectors. See Appendix B for information on how to enter values into any of the Vector Algebra Tools.

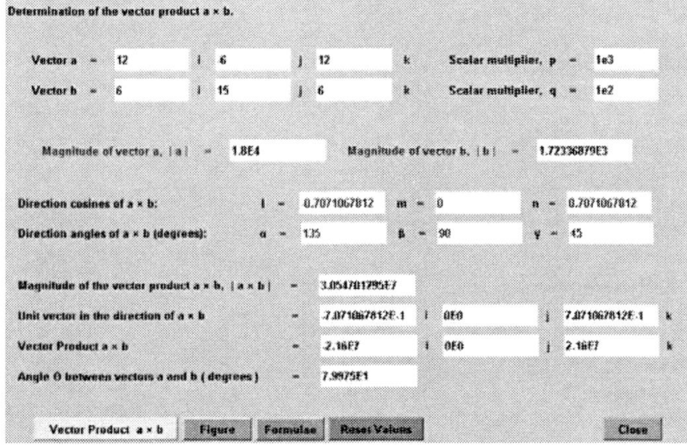

Fig. 5.10. The Vector Product Tool from the Vector Algebra Tools Software Program.

The **Forces on an Electric Charge Topic Tool** is used to determine the forces on an electric charge in an electric field, in a magnetic field or both. This vector topic tool enables you to calculate the following forces:

1. The force on a static or moving charge q in a uniform electric field $\mathbf{E}$, which is given by $\mathbf{F}_E = q\mathbf{E}$.

2. The force on a moving charge q in a uniform magnetic field $\mathbf{B}$, which is given by $\mathbf{F}_B = q(\mathbf{v} \times \mathbf{B})$, where $\mathbf{v}$ is the instantaneous velocity of the charge and $\mathbf{B}$ is the uniform magnetic field.

3. The **Lorentz** force, which is the net force acting on a moving charge q in both a uniform electric field $\mathbf{E}$ and a uniform magnetic field $\mathbf{B}$, and is given by $\mathbf{F}_L = \mathbf{F}_E + \mathbf{F}_B = q\mathbf{E} + q(\mathbf{v} \times \mathbf{B})$.

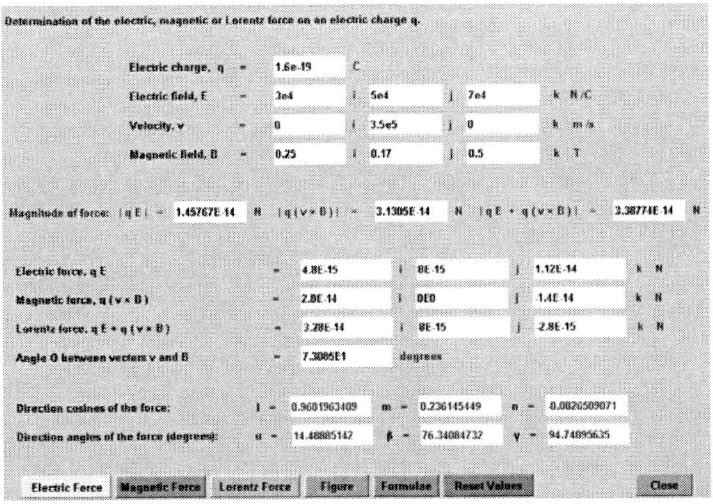

Fig. 5.11. The Forces on an Electric Charge Topic Tool from the Vector Algebra Tools Software Program.

To access the **Forces on an Electric Charge Topic Tool**, open the Vector Product menu, and select the **Forces on an Electric Charge Topic Tool**.

Question 1. Evaluate (a) $\mathbf{a} \times \mathbf{b}$ and (b) $\mathbf{b} \times \mathbf{a}$, given $\mathbf{a} = 2\,\mathbf{i} - 6\,\mathbf{j} + 2\,\mathbf{k}$ and $\mathbf{b} = 4\,\mathbf{i} - 7\,\mathbf{j} + 8\,\mathbf{k}$.

Worked Solutions

1(a) Solve for $\mathbf{a} \times \mathbf{b}$, where

$$\mathbf{a} \times \mathbf{b} = \begin{vmatrix} \mathbf{i} & \mathbf{j} & \mathbf{k} \\ a_x & a_y & a_z \\ b_x & b_y & b_z \end{vmatrix} = \begin{vmatrix} \mathbf{i} & \mathbf{j} & \mathbf{k} \\ 2 & -6 & 2 \\ 4 & -7 & 8 \end{vmatrix} = -34\,\mathbf{i} - 8\,\mathbf{j} + 10\,\mathbf{k}.$$

1(b) Solve for $\mathbf{b} \times \mathbf{a}$, where

$$\mathbf{b} \times \mathbf{a} = \begin{vmatrix} \mathbf{i} & \mathbf{j} & \mathbf{k} \\ b_x & b_y & b_z \\ a_x & a_y & a_z \end{vmatrix} = \begin{vmatrix} \mathbf{i} & \mathbf{j} & \mathbf{k} \\ 4 & -7 & 8 \\ 2 & -6 & 2 \end{vmatrix} = 34\,\mathbf{i} + 8\,\mathbf{j} - 10\,\mathbf{k}.$$

Since each vector product is normal to the plane containing the nonzero vectors **a** and **b**, you can check your answers by testing for the orthogonal condition between each vector that is in the plane and the corresponding vector products. Hence, using vector **a** as an example:

$$\mathbf{a} \bullet (\mathbf{a} \times \mathbf{b}) = (2\,\mathbf{i} - 6\,\mathbf{j} + 2\,\mathbf{k}) \bullet (-34\,\mathbf{i} - 8\,\mathbf{j} + 10\,\mathbf{k}) = 0 \text{ and}$$

$$\mathbf{a} \bullet (\mathbf{b} \times \mathbf{a}) = (4\,\mathbf{i} - 7\,\mathbf{j} + 8\,\mathbf{k}) \bullet (34\,\mathbf{i} + 8\,\mathbf{j} - 10\,\mathbf{k}) = 0.$$

Likewise, you can test for the orthogonal condition between vector **b** and the corresponding vector products $(\mathbf{a} \times \mathbf{b})$ and $(\mathbf{b} \times \mathbf{a})$.

Vector Algebra Tools Solutions

Using the Vector Product Tool: Enter the given components of vector **a** in the Vector **a** user input fields, and enter the given components of vector **b** in the Vector **b** user input fields. Then click the Vector Product $\mathbf{a} \times \mathbf{b}$ button. The answers will be displayed in the output fields as shown. Click the Reset Values button before continuing to the next step.

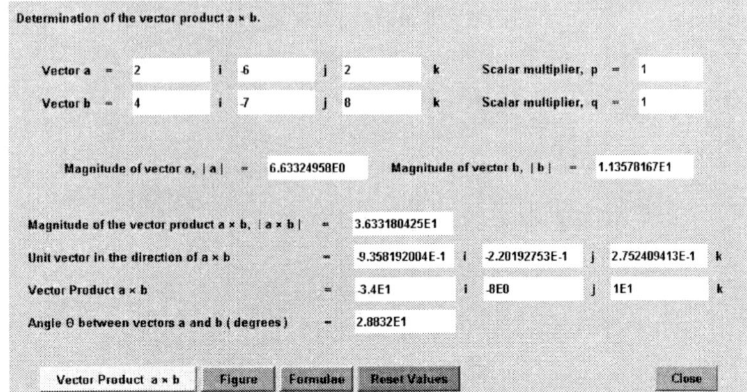

Solution: 1(a) $\mathbf{a} \times \mathbf{b} = -34\,\mathbf{i} - 8\,\mathbf{j} + 10\,\mathbf{k}$.

Using the Vector Product Tool: Enter the given components of vector **b** in the Vector **a** user input fields, and enter the given components of vector **a** in the Vector **b** user input fields. Then click the Vector Product **a** × **b** button. The answers will be displayed in the output fields as shown.

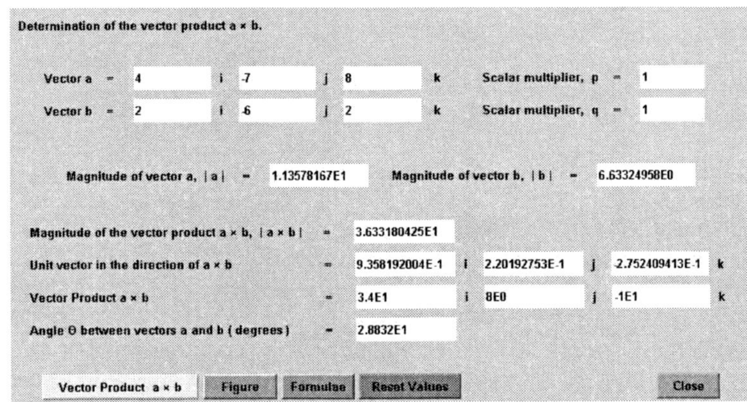

Solution: 1(b) $\mathbf{b} \times \mathbf{a} = 34\,\mathbf{i} + 8\,\mathbf{j} - 10\,\mathbf{k}$.

Answers: 1(a) $\mathbf{a} \times \mathbf{b} = -34\,\mathbf{i} - 8\,\mathbf{j} + 10\,\mathbf{k}$ and 1(b) $\mathbf{b} \times \mathbf{a} = 34\,\mathbf{i} + 8\,\mathbf{j} - 10\,\mathbf{k}$. Note that $\mathbf{a} \times \mathbf{b} = -(\mathbf{b} \times \mathbf{a})$, as the commutative law of vector algebra does not apply, but $|\,\mathbf{a} \times \mathbf{b}\,| = |\,\mathbf{b} \times \mathbf{a}\,|$.

Question 2. Determine (a) the instantaneous linear velocity $\mathbf{v}$ and (b) the angle θ between $\boldsymbol{\omega}$ and $\mathbf{r}$, given that $\mathbf{v} = \boldsymbol{\omega} \times \mathbf{r}$, where $\boldsymbol{\omega} = \mathbf{i} - 2\,\mathbf{j} + \mathbf{k}$ rad/s and $\mathbf{r} = 4\,\mathbf{i} + 4\,\mathbf{j} + 4\,\mathbf{k}$ m.

Worked Solutions

2(a) Solve for $\mathbf{v} = \boldsymbol{\omega} \times \mathbf{r}$, where

$$\mathbf{v} = \boldsymbol{\omega} \times \mathbf{r} = \begin{vmatrix} \mathbf{i} & \mathbf{j} & \mathbf{k} \\ \omega_x & \omega_y & \omega_z \\ r_x & r_y & r_z \end{vmatrix} = \begin{vmatrix} \mathbf{i} & \mathbf{j} & \mathbf{k} \\ 1 & -2 & 1 \\ 4 & 4 & 4 \end{vmatrix} = -12\,\mathbf{i} + 12\,\mathbf{k} \text{ m/s.}$$

2(b) Solve for the angle θ, where

$$\theta = \sin^{-1}\left(\frac{|\boldsymbol{\omega} \times \mathbf{r}|}{|\boldsymbol{\omega}|\,|\mathbf{r}|}\right)$$

$$= \sin^{-1}\left(\frac{16.9705627}{(2.44948974)\ (6.92820323)}\right) = \sin^{-1}(1.0) = 90.0°.$$

Vector Algebra Tools Solutions

Using the Vector Product Tool: Enter the given components of vector $\boldsymbol{\omega}$ in the Vector **a** user input fields, and enter the given components of vector **r** in the Vector **b** user input fields. Then click the Vector Product **a × b** button. The answers will be displayed in the output fields as shown.

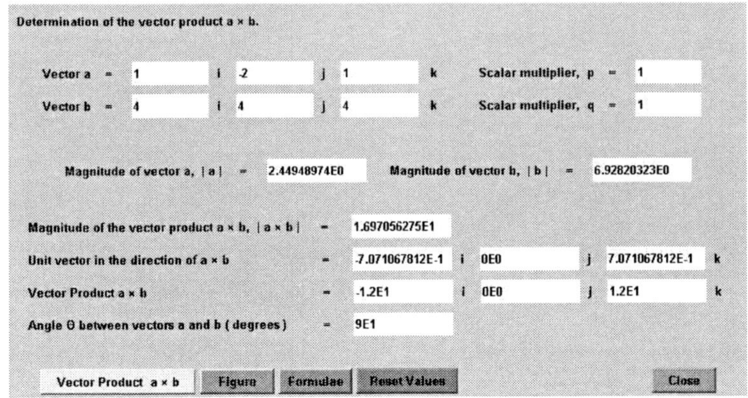

Answers: 2(a) $\mathbf{v} = -12\,\mathbf{i} + 12\,\mathbf{k}$ m/s and 2(b) $\theta = 90°$. Hence $\boldsymbol{\omega} \perp \mathbf{r}$.

Question 3. Determine the area of the triangle shown in figure 5.12 whose adjacent sides are given by $\mathbf{a} = \mathbf{i} + 2\,\mathbf{j} + 2\,\mathbf{k}$ and $\mathbf{b} = -4\,\mathbf{i} - 6\,\mathbf{j} - 4\,\mathbf{k}$. The unit length is the millimeter.

Fig. 5.12.

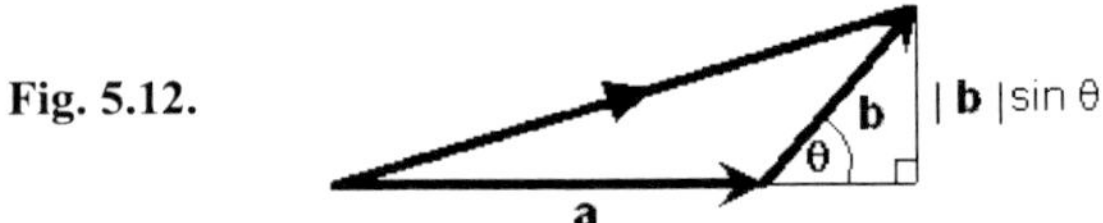

Worked Solution

The area Δ of the triangle is given by

$$\Delta = \tfrac{1}{2}\,|\,\mathbf{a}\,|\,|\,\mathbf{b}\,|\,\sin\theta = \tfrac{1}{2}\,|\,\mathbf{a} \times \mathbf{b}\,|.$$

Solve for the vector area **Δ** of the triangle, where

$$\boldsymbol{\Delta} = \tfrac{1}{2}\begin{vmatrix} \mathbf{i} & \mathbf{j} & \mathbf{k} \\ 1 & 2 & 2 \\ -4 & -6 & -4 \end{vmatrix}$$

$$= 5 \times 10^{-7}\,(4\,\mathbf{i} - 4\,\mathbf{j} + 2\,\mathbf{k}).$$

Therefore, the area of the triangle is given by

$$\Delta = 5 \times 10^{-7}\,|\,4\,\mathbf{i} - 4\,\mathbf{j} + 2\,\mathbf{k}\,|$$

$$= 5 \times 10^{-7}\,\sqrt{4^2 + (-4)^2 + 2^2}$$

$$= 3 \times 10^{-6}\ \text{m}^2.$$

Vector Algebra Tools Solution

Using the Vector Product Tool: Enter the given components of vector **a** expressed in meters in the Vector **a** user input fields, and enter the given components of vector **b** expressed in meters in the Vector **b** user input fields. Next enter the scalar multiplier of 0.5 in the Scalar multiplier, p user input field. Then click the Vector Product **a × b** button. The answers will be displayed in the output fields as shown.

Determination of the vector product a × b.

Vector a	=	0.001	i	0.002	j	0.002	k	Scalar multiplier, p =	0.5
Vector b	=	-0.004	i	-0.006	j	-0.004	k	Scalar multiplier, q =	1

Magnitude of vector a, | a | = 1.5E-3 Magnitude of vector b, | b | = 8.24621125E-3

Direction cosines of a × b:	l =	0.6666666667	m =	-0.6666666667	n =	0.3333333333
Direction angles of a × b (degrees):	α =	48.1896851	β =	131.8103149	γ =	70.52877937

Magnitude of the vector product a × b, | a × b |	=	3E-6					
Unit vector in the direction of a × b	=	6.666666667E-1	i	-6.666666667E-1	j	3.333333333E-1	k
Vector Product a × b	=	2E-6	i	-2E-6	j	1E-6	k
Angle Θ between vectors a and b (degrees)	=	1.4036E1					

Vector Product a × b | Figure | Formulae | Reset Values | Close

Answer: Area of the triangle, $\Delta = 3 \times 10^{-6}$ m^2.

Question 4. In a uniform external magnetic field **B**, it is observed that a positive electric charge of $q = 3.5$ μC moves with a speed of 525 m/s along the positive x-axis, as shown in figure 5.13. If the magnitude of **B** has a field strength of 2.5 T and points along the positive y-axis, determine the magnetic force vector $\mathbf{F}_B$, its magnitude, $|\mathbf{F}_B|$, and the direction of the force acting on q. Also, if a negative charge is substituted for the positive charge, what is the direction of the magnetic force?

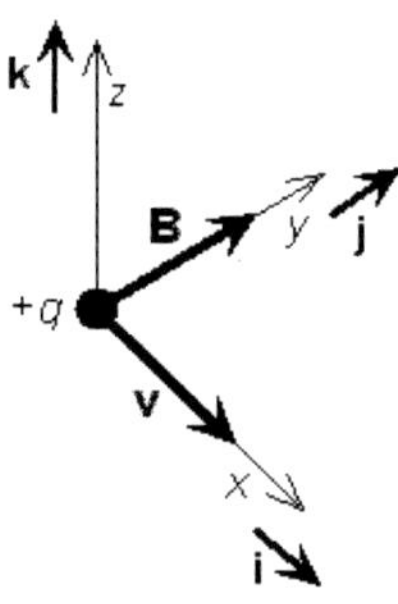

Fig. 5.13.

Worked Solutions

The instantaneous velocity vector of the positive charge is $\mathbf{v} = 525\,\mathbf{i}$ m/s and the external magnetic field vector is $\mathbf{B} = 2.5\,\mathbf{j}$ T.

Solve for the magnetic force vector $\mathbf{F}_B$, where

$$\mathbf{F}_B = q(\mathbf{v} \times \mathbf{B})$$

$$= 3.5 \times 10^{-6} \begin{vmatrix} \mathbf{i} & \mathbf{j} & \mathbf{k} \\ v_x & v_y & v_z \\ B_x & B_y & B_z \end{vmatrix}$$

$$= 3.5 \times 10^{-6} \begin{vmatrix} \mathbf{i} & \mathbf{j} & \mathbf{k} \\ 525 & 0 & 0 \\ 0 & 2.5 & 0 \end{vmatrix}$$

$$= 4.5938 \times 10^{-3} \, \mathbf{k} \text{ N}.$$

Clearly, $\mathbf{F}_B$ points along the positive z-axis in the direction of the **k** unit vector and is normal to the plane containing the vectors **v** and **B**.

The magnitude of $\mathbf{F}_B$, $|\,\mathbf{F}_B\,| = |\,4.5938 \times 10^{-3}\,\mathbf{k}\,| = 4.5938 \times 10^{-3}$ N.

If the electric charge q is negative, then the direction of $\mathbf{F}_B$ is given by $\mathbf{F}_B = -q(\mathbf{v} \times \mathbf{B})$ and is along the negative z-axis in the direction of the $-\mathbf{k}$ unit vector.

Vector Algebra Tools Solutions

Using the Vector Product Tool: Enter the given components of vector **v** in the Vector **a** user input fields, and enter the given components of vector **B** in the Vector **b** user input fields. Next enter the scalar multiplier of 3.5×10^{-6} in scientific notation in the Scalar multiplier, p user input field. Then click the Vector Product **a × b** button. The answers will be displayed in the output fields as shown.

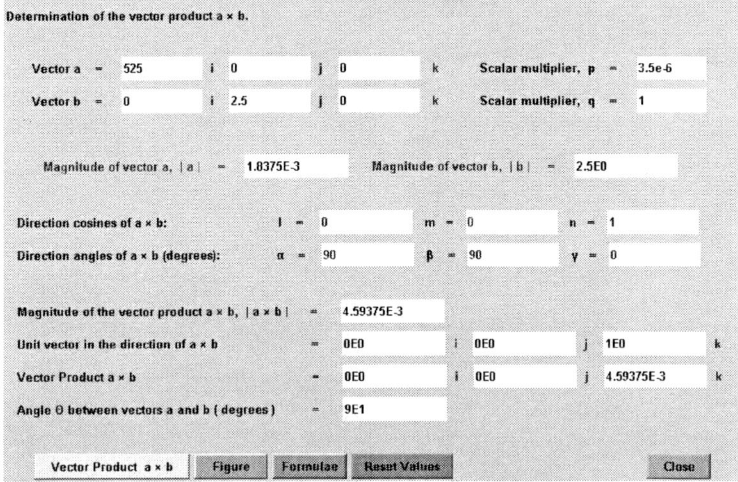

Answers: $\mathbf{F}_B = 4.5938 \times 10^{-3}\,\mathbf{k}$ N and $|\,\mathbf{F}_B\,| = 4.5938 \times 10^{-3}$ N. The direction of $\mathbf{F}_B$ for a positive charge is along the positive z-axis, normal to the plane containing vectors **v** and **B** and is in the direction of the **k** unit vector. For a negative charge, the direction of $\mathbf{F}_B$ is along the negative z-axis in the direction of the $-\mathbf{k}$ unit vector.

Note that question 4 can also be solved using the **Forces on an Electric Charge Topic Tool** as setout below.

Using the Forces on an Electric Charge Topic Tool: Enter the given value of the electric charge q in the Electric charge, q input field; the components of the velocity **v** in the Velocity, **v** user input fields and the components of the magnetic field **B** in the Magnetic field, **B** user input fields. Then click the Magnetic Force button. The answers will be displayed in the output fields as shown.

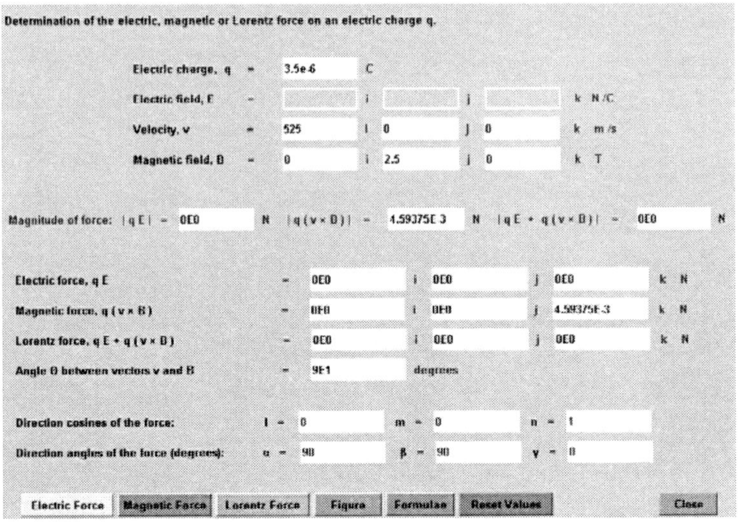

Determination of the electric, magnetic or Lorentz force on an electric charge q.

Electric charge, q	=	3.5e-6	C			
Electric field, E	=		i		j	k N/C
Velocity, v	=	525	i 0		j 0	k m/s
Magnetic field, B	=	0	i 2.5		j 0	k T

Magnitude of force: \| q E \| = 0E0 N \| q (v × B) \| = 4.59375E-3 N \| q E + q (v × B) \| = 0E0 N

Electric force, q E	=	0E0	i 0E0	j 0E0	k N
Magnetic force, q (v × B)	=	0E0	i 0E0	j 4.59375E-3	k N
Lorentz force, q E + q (v × B)	=	0E0	i 0E0	j 0E0	k N
Angle θ between vectors v and B	=	9E1	degrees		

Direction cosines of the force:	l = 0	m = 0	n = 1
Direction angles of the force (degrees):	α = 90	β = 90	γ = 0

Electric Force | Magnetic Force | Lorentz Force | Figure | Formulae | Reset Values | Close

Answers: $\mathbf{F}_B = 4.59375 \times 10^{-3}$ **k** N and $|\mathbf{F}_B| = 4.59375 \times 10^{-3}$ N. The direction of $\mathbf{F}_B$ for a positive charge is along the positive z-axis, normal to the plane containing vectors **v** and **B** and in the direction of the **k** unit vector. For a negative charge, the direction of $\mathbf{F}_B$ is along the negative z-axis in the direction of the $-\mathbf{k}$ unit vector.

Question 5 (a) Determine the magnitude and direction cosines of the electric force $\mathbf{F}_E$ that a uniform electric field $\mathbf{E} = 10^4\ (9\ \mathbf{i} + 4\ \mathbf{j} - 5\ \mathbf{k})$ N/C exerts on a stationary electric charge $q = 3\ \mu$C.

Worked Solutions

Solve for the electric force $\mathbf{F}_E$, where

$$\mathbf{F}_E = q\mathbf{E} = 3 \times 10^{-6}\,[10^4\ (9\ \mathbf{i} + 4\ \mathbf{j} - 5\ \mathbf{k})] = 0.27\ \mathbf{i} + 0.12\ \mathbf{j} - 0.15\ \mathbf{k}\ \text{N}.$$

The magnitude of the electric force $\mathbf{F}_E$ is given by

$$|\mathbf{F}_E| = |q\mathbf{E}| = \sqrt{0.27^2 + 0.12^2 + (-0.15)^2} = 0.33136\ \text{N}.$$

The direction cosines of $\mathbf{F}_E$ are given by

$$\ell = \cos\alpha = \frac{F_x}{|\mathbf{F}_E|} = \frac{0.27}{0.33136} = 0.8148,$$

$$m = \cos\beta = \frac{F_y}{|\,\mathbf{F}_\mathrm{E}\,|} = \frac{0.12}{0.33136} = 0.3621 \text{ and}$$

$$n = \cos\gamma = \frac{F_z}{|\,\mathbf{F}_\mathrm{E}\,|} = \frac{-0.15}{0.33136} = -0.4527.$$

Vector Algebra Tools Solutions

Using the Forces on an Electric Charge Topic Tool: Enter the given value of the electric charge q in the Electric charge, q input field and the components of the electric field **E** in the Electric field, **E** user input fields. Then click the Electric Force button. The answers will be displayed in the output fields as shown.

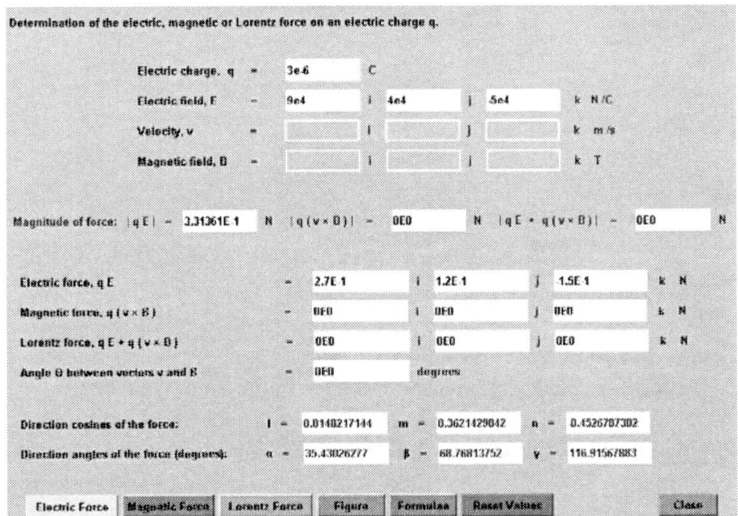

Answers: $|\,\mathbf{F}_\mathrm{E}\,| = |\,q\mathbf{E}\,| = 0.33136$ N and $\ell = 0.8148$, $m = 0.3621$ and $n = -0.4527$. Note that you can also use the Vector Arithmetic Tool to solve for $|\,\mathbf{F}_\mathrm{E}\,|$.

Question 5 (b) If the electric charge $q = 3.5\ \mu C$ moves with an instantaneous velocity $\mathbf{v} = 2 \times 10^5\,\mathbf{j} + 5.5 \times 10^5\,\mathbf{k}$ m/s through a uniform magnetic field $\mathbf{B} = 1.5\,\mathbf{i} - 0.75\,\mathbf{j} + 0.25\,\mathbf{k}$ T, determine the magnitude of the magnetic force $\mathbf{F}_B$ on q.

Worked Solution

Solve for $\mathbf{F}_B$ where

$$\mathbf{F}_B = q(\mathbf{v} \times \mathbf{B}) = 3 \times 10^{-6} \begin{vmatrix} \mathbf{i} & \mathbf{j} & \mathbf{k} \\ v_x & v_y & v_z \\ B_x & B_y & B_z \end{vmatrix}$$

$$= 3 \times 10^{-6} \begin{vmatrix} \mathbf{i} & \mathbf{j} & \mathbf{k} \\ 0 & 2\times10^5 & 5.5\times10^5 \\ 1.5 & -0.75 & 0.25 \end{vmatrix}$$

$$= 1.3875\,\mathbf{i} + 2.475\,\mathbf{j} - 0.9\,\mathbf{k}\ \text{N}.$$

The magnitude of the magnetic force $\mathbf{F}_B$ is given by

$$|\,\mathbf{F}_B\,| = |\,q(\mathbf{v} \times \mathbf{B})\,| = \sqrt{1.3875^2 + 2.475^2 + (-0.9)^2} = 2.9767\ \text{N}.$$

Vector Algebra Tools Solution

Using the Forces on an Electric Charge Topic Tool: Enter the given value of the electric charge in the Electric charge, q input field, the components of the velocity **v** in the Velocity, **v** user input fields, and the components of the magnetic field **B** in the Magnetic field, **B** user input fields. Then click the Magnetic Force button. The answers will be displayed in the output fields as shown.

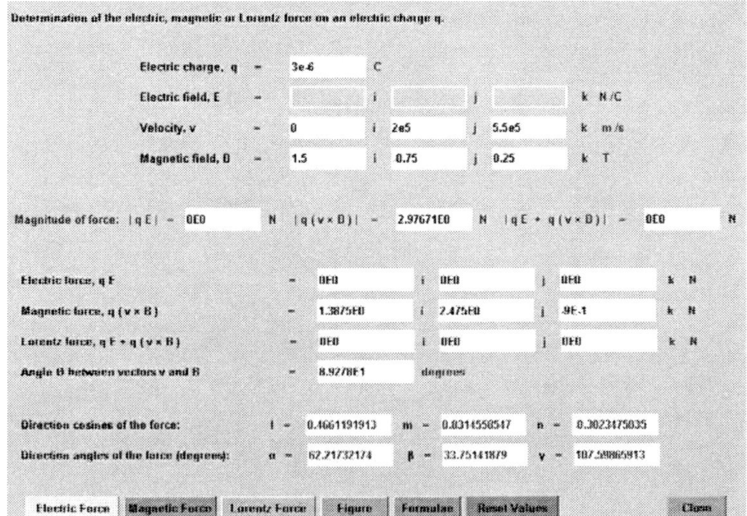

Answer: $|\mathbf{F}_B| = |q(\mathbf{v} \times \mathbf{B})| = 2.9767$ N. Note that you can also use the Vector Product Tool to solve for $|\mathbf{F}_B|$.

Question 5 (c). Determine the vector components and corresponding magnitude of the Lorentz force that is exerted on an electric charge $q = 3.5\ \mu$C that moves with an instantaneous velocity $\mathbf{v} = 2 \times 10^5\ \mathbf{j} + 5.5 \times 10^5\ \mathbf{k}$ m/s through a uniform electric field $\mathbf{E} = 10^4\ (9\ \mathbf{i} + 4\ \mathbf{j} - 5\ \mathbf{k})$ N/C and magnetic field $\mathbf{B} = 1.5\ \mathbf{i} - 0.75\ \mathbf{j} + 0.25\ \mathbf{k}$ T.

Worked Solutions

The Lorentz force $\mathbf{F}_L$ on the charge q is given by the vector sum of $\mathbf{F}_E$ and $\mathbf{F}_B$, where

$$\mathbf{F}_L = \mathbf{F}_E + \mathbf{F}_B$$

$$= (0.27\ \mathbf{i} + 0.12\ \mathbf{j} - 0.15\ \mathbf{k}) + (1.3875\ \mathbf{i} + 2.475\ \mathbf{j} - 0.9\ \mathbf{k})$$

$$= 1.6575\ \mathbf{i} + 2.595\ \mathbf{j} - 1.05\ \mathbf{k}\ \text{N}.$$

The magnitude of the net force $\mathbf{F}_L$ is given by

$$|\mathbf{F}_L| = \sqrt{1.6575^2 + 2.595^2 + (-1.05)^2} = 3.2533\ \text{N}.$$

Vector Algebra Tools Solutions

Using the Forces on an Electric Charge Topic Tool: Enter the given value of the electric charge q in the Electric charge, q input field, the components of the velocity **v** in the Velocity, **v** user input fields, the components of the electric field **E** in the Electric field, **E** user input fields, and the components of the magnetic field **B** in the Magnetic field, **B** user input fields. Then click the Lorentz Force button. The answers will be displayed in the output fields as shown.

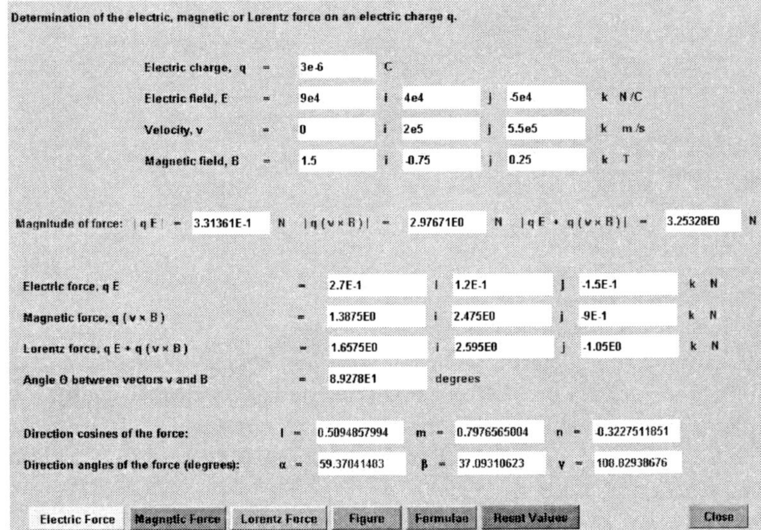

Answers: $\mathbf{F}_L = 1.6575\,\mathbf{i} + 2.595\,\mathbf{j} - 1.05\,\mathbf{k}$ N and $|\mathbf{F}_L| = 3.2533$ N. Note that you can also use the Vector Arithmetic Tool to solve for $\mathbf{F}_L$ and $|\mathbf{F}_L|$, since $\mathbf{F}_L$ is the vector sum of $\mathbf{F}_E$ and $\mathbf{F}_B$.

5.7 NOTES:

CHAPTER 6

SCALAR AND VECTOR TRIPLE PRODUCTS

In this chapter the triple products of three vectors, which results in either a scalar or a vector quantity, are presented. In particular, scalar and vector triple product definitions, properties, problem solving applications and worked examples are included. Also demonstrated is the use of the Triple Products software tool in solving the example problems.

6.1 DEFINITIONS AND DETERMINATIONS OF SCALAR AND VECTOR TRIPLE PRODUCT QUANTITIES

6.1.1 Scalar triple product

A **scalar triple product** or **box product** is the scalar quantity resulting from the scalar product of vector **a** and the vector (**b** × **c**) and is given by

$$\mathbf{a} \bullet (\mathbf{b} \times \mathbf{c}).$$

Since the vector product operation takes precedence over the scalar product operation, the vector product is always calculated first. Therefore, the parentheses may be removed in the scalar triple product expression for convenience, and the scalar triple product may be expressed simply as $\mathbf{a} \bullet \mathbf{b} \times \mathbf{c}$.

If vectors **a**, **b** and **c** are expressed in their component form in a three-dimensional Cartesian coordinate system, where $\mathbf{a} = a_x\,\mathbf{i} + a_y\,\mathbf{j} + a_z\,\mathbf{k}$, $\mathbf{b} = b_x\,\mathbf{i} + b_y\,\mathbf{j} + b_z\,\mathbf{k}$ and $\mathbf{c} = c_x\,\mathbf{i} + c_y\,\mathbf{j} + c_z\,\mathbf{k}$, then the scalar triple product $\mathbf{a} \bullet (\mathbf{b} \times \mathbf{c})$ can be determined algebraically as follows:

$$\mathbf{a} \bullet (\mathbf{b} \times \mathbf{c}) = (a_x\,\mathbf{i} + a_y\,\mathbf{j} + a_z\,\mathbf{k}) \bullet [\,(b_y c_z - b_z c_y)\,\mathbf{i} + (b_z c_x - b_x c_z)\,\mathbf{j} + (b_x c_y - b_y c_x)\,\mathbf{k}\,]$$

$$= a_x (b_y c_z - b_z c_y) + a_y (b_z c_x - b_x c_z) + a_z (b_x c_y - b_y c_x).$$

This expression is more commonly written in determinant form, where the components of the vectors **a**, **b** and **c** correspond to the three rows of a 3 × 3 determinant, such that

$$\mathbf{a} \bullet (\mathbf{b} \times \mathbf{c}) = \begin{vmatrix} a_x & a_y & a_z \\ b_x & b_y & b_z \\ c_x & c_y & c_z \end{vmatrix}.$$

Based upon the properties of determinants, the determinant of a transposed matrix has the same value as the original. Therefore, the components of the vectors **a**, **b** and **c** can also be written as columns in the determinant, such that

$$\mathbf{a} \bullet (\mathbf{b} \times \mathbf{c}) = \begin{vmatrix} a_x & b_x & c_x \\ a_y & b_y & c_y \\ a_z & b_z & c_z \end{vmatrix}.$$

The order of the two vectors, vector **a** and the vector (**b** × **c**), in the scalar triple product equation is not important and, therefore, may be reversed. Thus,

$$\mathbf{a} \bullet (\mathbf{b} \times \mathbf{c}) = (\mathbf{b} \times \mathbf{c}) \bullet \mathbf{a}.$$

Also, by interchanging the dot and cross product symbols in the scalar triple product, the value of the scalar triple product is unchanged. For example,

$$(\mathbf{a} \times \mathbf{b}) \bullet \mathbf{c} = \mathbf{b} \bullet (\mathbf{c} \times \mathbf{a}).$$

It also follows from the properties of determinants that the value of the scalar triple product is invariant (unchanged) by the cyclical interchange of the three rows (or columns) in the determinant, such that

$$\begin{vmatrix} a_x & a_y & a_z \\ b_x & b_y & b_z \\ c_x & c_y & c_z \end{vmatrix} = \begin{vmatrix} b_x & b_y & b_z \\ c_x & c_y & c_z \\ a_x & a_y & a_z \end{vmatrix} = \begin{vmatrix} c_x & c_y & c_z \\ a_x & a_y & a_z \\ b_x & b_y & b_z \end{vmatrix}.$$

These determinants, in their respective order, are equivalent to the following scalar triple product identities:

$$\mathbf{a} \bullet (\mathbf{b} \times \mathbf{c}) = \mathbf{b} \bullet (\mathbf{c} \times \mathbf{a}) = \mathbf{c} \bullet (\mathbf{a} \times \mathbf{b}).$$

However, the exchange of only two of the three vectors in the scalar triple product calculation will result in the change of the sign in the answer. For example,

$$\begin{aligned}\mathbf{a} \bullet (\mathbf{b} \times \mathbf{c}) &= -\mathbf{c} \bullet (\mathbf{b} \times \mathbf{a}) \\ \mathbf{b} \bullet (\mathbf{c} \times \mathbf{a}) &= -\mathbf{a} \bullet (\mathbf{c} \times \mathbf{b}) \\ \mathbf{c} \bullet (\mathbf{a} \times \mathbf{b}) &= -\mathbf{b} \bullet (\mathbf{a} \times \mathbf{c}).\end{aligned}$$

6.1.2 Vector triple products

Vector triple products are vector quantities resulting from the vector product of vector **a** and the vector (**b** × **c**) or from the vector product of the vector (**a** × **b**) and vector **c**. These two vector triple products are expressed as

$$\mathbf{a} \times (\mathbf{b} \times \mathbf{c}) \text{ and } (\mathbf{a} \times \mathbf{b}) \times \mathbf{c}.$$

Since only vector products are involved, and because these operations are neither commutative nor associative, the order in which the vector product calculations are performed is important. Hence, the parentheses are necessary, as

$$\mathbf{a} \times (\mathbf{b} \times \mathbf{c}) \neq (\mathbf{a} \times \mathbf{b}) \times \mathbf{c}.$$

For example, given that **a** = **i**, **b** = **i** and **c** = **k**, we have the following vector triple product results:

$$\mathbf{a} \times (\mathbf{b} \times \mathbf{c}) = \mathbf{i} \times (\mathbf{i} \times \mathbf{k}) = \mathbf{i} \times -\mathbf{j} = -\mathbf{k} \text{ and}$$

$$(\mathbf{a} \times \mathbf{b}) \times \mathbf{c} = (\mathbf{i} \times \mathbf{i}) \times \mathbf{k} = \mathbf{0} \times \mathbf{k} = \mathbf{0}.$$

Clearly, the above results are quite different depending on the order in which the vector products are performed.

The vector triple products can be expressed in determinant form as follows:

$$\mathbf{a} \times (\mathbf{b} \times \mathbf{c}) = \begin{vmatrix} \mathbf{i} & \mathbf{j} & \mathbf{k} \\ a_x & a_y & a_z \\ b_y c_z - b_z c_y & b_z c_x - b_x c_z & b_x c_y - b_y c_x \end{vmatrix}.$$

$$(\mathbf{a} \times \mathbf{b}) \times \mathbf{c} = \begin{vmatrix} \mathbf{i} & \mathbf{j} & \mathbf{k} \\ a_y b_z - a_z b_y & a_z b_x - a_x b_z & a_x b_y - a_y b_x \\ c_x & c_y & c_z \end{vmatrix}.$$

An alternative and simpler method of calculating the vector triple products is to express the vector triple product in the form of its vector triple product identity. For instance, figure 6.1 shows the geometrical representation of the vector triple product $\mathbf{a} \times (\mathbf{b} \times \mathbf{c})$, where the vector $\mathbf{a} \times (\mathbf{b} \times \mathbf{c})$ is a coplanar vector that lies in the same plane as the noncollinear vectors **b** and **c**. Therefore, the vector triple product $\mathbf{a} \times (\mathbf{b} \times \mathbf{c})$ can be expressed as a linear combination of the vectors **b** and **c**, such that

$$\mathbf{a} \times (\mathbf{b} \times \mathbf{c}) = p\mathbf{b} + q\mathbf{c},$$

where the scalar constants are given by $p = \mathbf{a} \bullet \mathbf{c}$ and $q = -\mathbf{a} \bullet \mathbf{b}$. Consequently, $\mathbf{a} \times (\mathbf{b} \times \mathbf{c})$ can be represented by the vector triple product identity

$$\mathbf{a} \times (\mathbf{b} \times \mathbf{c}) = \mathbf{b}(\mathbf{a} \bullet \mathbf{c}) - \mathbf{c}(\mathbf{a} \bullet \mathbf{b}).$$

This relationship is known as the "BAC – CAB" rule.

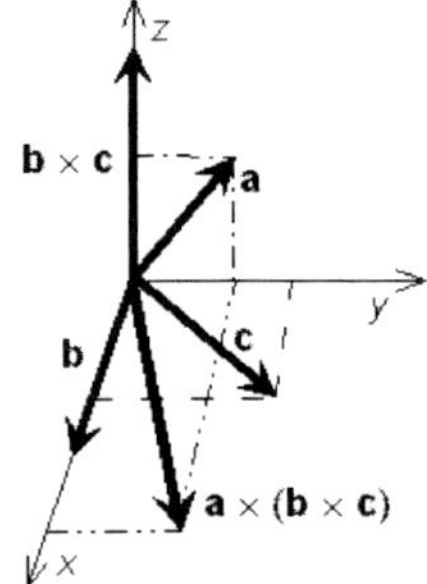

Fig. 6.1. The vector triple product $\mathbf{a} \times (\mathbf{b} \times \mathbf{c})$ is a vector that lies in the same plane as the noncollinear vectors **b** and **c**.

Likewise, the vector $(\mathbf{a} \times \mathbf{b}) \times \mathbf{c}$ is a vector that lies in the same plane as the noncollinear vectors **a** and **b** and can be expressed as the linear combination of the vectors **a** and **b**, such that $p\mathbf{a} + q\mathbf{b}$, where the scalar constants are given by $p = -\mathbf{b} \bullet \mathbf{c}$ and $q = \mathbf{a} \bullet \mathbf{c}$. Consequently, $(\mathbf{a} \times \mathbf{b}) \times \mathbf{c}$ can be represented by the vector triple product identity

$$(\mathbf{a} \times \mathbf{b}) \times \mathbf{c} = \mathbf{b}(\mathbf{a} \bullet \mathbf{c}) - \mathbf{a}(\mathbf{b} \bullet \mathbf{c}).$$

This relationship is known as the "BAC – ABC" rule.

6.2 IMPORTANT RESULTS INVOLVING SCALAR AND VECTOR TRIPLE PRODUCTS

Important results and operations involving triple vector products are outlined as follows:

- The volume of a parallelepiped[1], V, whose adjacent sides are defined by the vectors **a**, **b** and **c**, is determined by the absolute value of the scalar triple product $|\mathbf{a} \bullet (\mathbf{b} \times \mathbf{c})|$ $(= |\mathbf{b} \bullet (\mathbf{c} \times \mathbf{a})| = |\mathbf{c} \bullet (\mathbf{a} \times \mathbf{b})|)$. This is illustrated in figure 6.2, where the base of the parallelepiped is the parallelogram defined by the vectors **b** and **c** of area $|\mathbf{b} \times \mathbf{c}|$. The vertical height is the scalar projection of vector **a** in the direction of the unit vector $\hat{\mathbf{n}}$ that is normal to the plane of the parallelogram, and is given by $\mathbf{a} \bullet \hat{\mathbf{n}} = |\mathbf{a}| \cos \phi$, where ϕ in the angle between $\mathbf{b} \times \mathbf{c}$ and **a**. Hence, the volume of the parallelepiped is equal to the area of its base times its vertical height and is given by

$$V = |\mathbf{b} \times \mathbf{c}|\,|\mathbf{a}| \cos \phi$$

$$= |(\mathbf{b} \times \mathbf{c}) \bullet \mathbf{a}|.$$

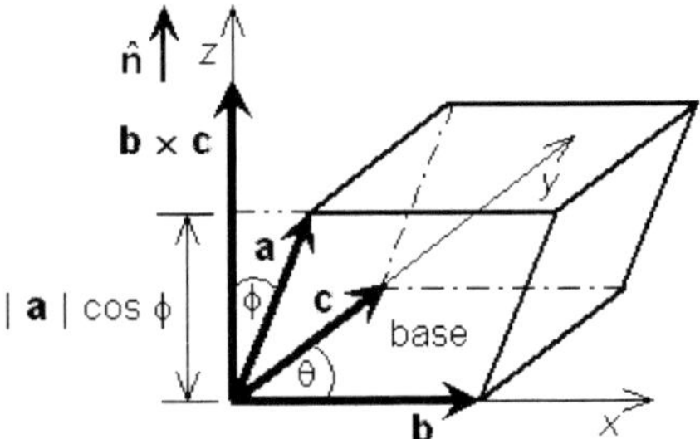

Fig. 6.2. Volume of a parallelepiped, whose sides are defined by the vectors **a**, **b** and **c**, is given by the expression $|\mathbf{b} \times \mathbf{c}|\,|\mathbf{a}| \cos \phi$, where $0° \leq \phi < 90°$.

When $\mathbf{a} \bullet \hat{\mathbf{n}} > 0$, such that the angle ϕ is acute, the scalar triple product is a positive number. In contrast, when $\mathbf{a} \bullet \hat{\mathbf{n}} < 0$, such that the angle ϕ is obtuse, the scalar triple product is a negative number. Note that the absolute value of the scalar triple product is required in this case, since the vectors **a**, **b** and **c** do not form a right-handed coordinate system.

---

[1] A parallelepiped is a six-faced polyhedron in which each face is a parallelogram and the opposite pairs of faces are congruent.

- If the scalar triple product $\mathbf{a} \cdot (\mathbf{b} \times \mathbf{c}) = 0$, then the three nonzero vectors **a**, **b** and **c** are coplanar and the volume of the parallelepiped is zero. Specifically, the vector $(\mathbf{b} \times \mathbf{c})$ is normal to the plane of **b** and **c** and is also normal to vector **a**. Hence, a common test to verify that three vectors are coplanar is a determination of the scalar triple product, such that $\mathbf{a} \cdot (\mathbf{b} \times \mathbf{c}) = 0$.

- If two of the three vectors, **a**, **b** and **c** are equal, then the scalar triple product of the three vectors is zero. For example,

$$(\mathbf{a} \times \mathbf{b}) \cdot \mathbf{a} = (\mathbf{a} \times \mathbf{b}) \cdot \mathbf{b} = (\mathbf{a} \times \mathbf{a}) \cdot \mathbf{c} = 0.$$

  Furthermore, if two of the three vectors are collinear, such that $\mathbf{b} = \lambda\mathbf{c}$, then

$$\mathbf{a} \cdot (\mathbf{b} \times \mathbf{c}) = \mathbf{a} \cdot (\lambda\mathbf{c} \times \mathbf{c}) = 0.$$

- If **i**, **j** and **k** are the unit vectors that are along the respective mutually perpendicular positive x, y and z-axes of a three-dimensional Cartesian coordinate system, then

$$\mathbf{i} \cdot (\mathbf{j} \times \mathbf{k}) = \mathbf{j} \cdot (\mathbf{k} \times \mathbf{i}) = \mathbf{k} \cdot (\mathbf{i} \times \mathbf{j}) = 1.$$

- If vectors **a**, **b**, **c** and **d** are nonzero vectors, then

$$\begin{aligned}(\mathbf{a} \times \mathbf{b}) \times (\mathbf{c} \times \mathbf{d}) &= (\mathbf{a} \times \mathbf{b} \cdot \mathbf{d})\mathbf{c} - (\mathbf{a} \times \mathbf{b} \cdot \mathbf{c})\mathbf{d} \\ &= (\mathbf{a} \cdot \mathbf{b} \times \mathbf{d})\mathbf{c} - (\mathbf{a} \cdot \mathbf{b} \times \mathbf{c})\mathbf{d} \\ &= (\mathbf{a} \cdot \mathbf{c} \times \mathbf{d})\mathbf{b} - (\mathbf{b} \cdot \mathbf{c} \times \mathbf{d})\mathbf{a}\end{aligned}$$

- If vectors **a**, **b**, **c** and **d** are nonzero vectors, where **a** is parallel to **b** and **c** is parallel to **d**, then

$$(\mathbf{a} \times \mathbf{b}) \times (\mathbf{c} \times \mathbf{d}) = \mathbf{0}.$$

  Geometrically, all four vectors are coplanar.

- If vectors **a**, **b** and **c** are nonzero vectors, then the Jacobi vector identity is given by

$$\mathbf{a} \times (\mathbf{b} \times \mathbf{c}) + \mathbf{b} \times (\mathbf{c} \times \mathbf{a}) + \mathbf{c} \times (\mathbf{a} \times \mathbf{b}) = \mathbf{0}.$$

6.3 PROBLEM SOLVING APPLICATIONS OF SCALAR AND VECTOR TRIPLE PRODUCTS

The following examples demonstrate various problem solving applications of scalar and vector triple products:

- In crystallography, the determination of the volume, V, of a crystallographic unit cell whose adjacent sides are defined by the vectors **a**, **b** and **c**, and is given by

$$V = |\mathbf{a} \bullet (\mathbf{b} \times \mathbf{c})|.$$

- The condition that two straight lines represented by the vector equations $\mathbf{r} = \mathbf{a}_1 + \lambda\mathbf{b}_1$ and $\mathbf{r} = \mathbf{a}_2 + \mu\mathbf{b}_2$, where $\mathbf{b}_1$ and $\mathbf{b}_2$ are nonparallel direction vectors intersect, is given by the scalar triple product

$$(\mathbf{a}_1 - \mathbf{a}_2) \bullet (\mathbf{b}_1 \times \mathbf{b}_2) = 0.$$

  For the two lines to intersect they must be coplanar such that the vector normal to their common plane is perpendicular to both lines and the line joining the points with the position vectors $\mathbf{a}_1$ and $\mathbf{a}_2$.

- Two nonintersecting straight lines expressed by their vector equations $\mathbf{r} = \mathbf{a}_1 + \lambda\mathbf{b}_1$ and $\mathbf{r} = \mathbf{a}_2 + \mu\,\mathbf{b}_2$ have a common normal that is parallel to the vector $\mathbf{b}_1 \times \mathbf{b}_2$. Therefore, the length of the common normal d, i.e., the shortest distance between the two lines, is the scalar projection of the line joining the points $\mathbf{a}_1$ and $\mathbf{a}_2$ in the direction of the common normal, and is given by the expression

$$d = \frac{|(\mathbf{a}_1 - \mathbf{a}_2) \bullet (\mathbf{b}_1 \times \mathbf{b}_2)|}{|\mathbf{b}_1 \times \mathbf{b}_2|}.$$

- The condition that four points, A, B, C and D, whose position vectors are **a**, **b**, **c** and **d**, respectively, are coplanar if it can be shown that the volume of a parallelepiped, whose sides are given by the vectors (**b** − **a**), (**c** − **a**) and (**d** − **a**) is zero, such that

$$(\mathbf{b} - \mathbf{a}) \bullet [(\mathbf{c} - \mathbf{a}) \times (\mathbf{d} - \mathbf{a})] = 0.$$

- The volume of a tetrahedron whose adjacent sides are defined by the vectors **a**, **b** and **c**, is given by $\frac{1}{6}\,|\,\mathbf{a} \bullet (\mathbf{b} \times \mathbf{c})\,|$. Note that a tetrahedron is a four-faced polyhedron with each face a triangle.

- The condition that the vectors **a**, **b** and **c**, which are given by $\mathbf{a} = a_x\,\mathbf{i} + a_y\,\mathbf{j} + a_z\,\mathbf{k}$, $\mathbf{b} = b_x\,\mathbf{i} + b_y\,\mathbf{j} + b_z\,\mathbf{k}$ and $\mathbf{c} = c_x\,\mathbf{i} + c_y\,\mathbf{j} + c_z\,\mathbf{k}$, are linearly independent if it can be shown that the determinant formed from the components of each vector is not equal to zero. Therefore, vectors **a**, **b** and **c** are linearly independent given that

$$\begin{vmatrix} a_x & a_y & a_z \\ b_x & b_y & b_z \\ c_x & c_y & c_z \end{vmatrix} \neq 0, \quad \text{i.e., } \mathbf{a} \bullet (\mathbf{b} \times \mathbf{c}) \neq 0,$$

  and are linearly dependent otherwise. This condition is satisfied if and only if **a**, **b** and **c** are nonzero and noncoplanar vectors.

- The orthogonal projection of the area of a parallelogram with adjacent sides defined by the vectors **a** and **b** onto a plane whose direction normal to the plane is defined by the unit vector $\hat{\mathbf{n}}$, is given by $|\,(\mathbf{a} \times \mathbf{b}) \bullet \hat{\mathbf{n}}\,|$.

- The moment of a force **F** about a line that passes through the point O parallel to a unit vector $\hat{\mathbf{n}}$ is the scalar projection of the vector moment of the force **F** about O in the direction of the line and is equal to the scalar quantity $(\mathbf{r} \times \mathbf{F}) \bullet \hat{\mathbf{n}}$, where **r** is the position vector of any point S on the line of action of the force, **F**, with respect to O. See figure 6.3.

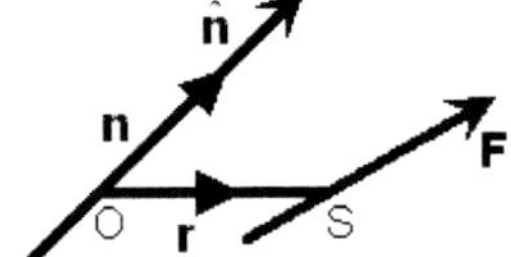

Fig. 6.3. Moment of a force **F** about a nonintersecting line through a point O is given by $(\mathbf{r} \times \mathbf{F}) \bullet \hat{\mathbf{n}}$.

6.4 WORKED EXAMPLES

The **Triple Products Tool** can be used can be used to determine the scalar and vector triple products of three vectors in component form in a Cartesian coordinate system. In addition, the tool can also be used to calculate the magnitude of the vector triple products. See Appendix B for information on how to enter values into any of the Vector Algebra Tools.

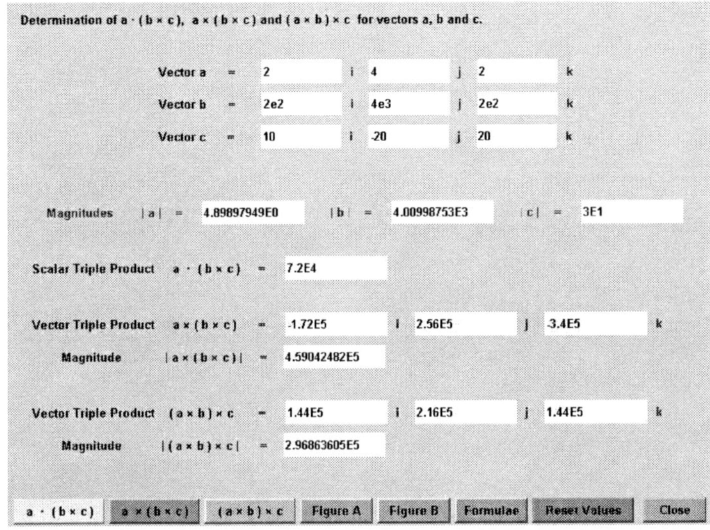

Fig. 6.4. The Triple Products Tool from the Vector Algebra Tools Software Program.

Question 1. If $\mathbf{a} = 4\,\mathbf{i} - 3\,\mathbf{j} + 6\,\mathbf{k}$ m, $\mathbf{b} = 2\,\mathbf{i} + 5\,\mathbf{j} + 2\,\mathbf{k}$ m and $\mathbf{c} = 3\,\mathbf{i} - \mathbf{j} + 7\,\mathbf{k}$ m, calculate (a) $\mathbf{a} \bullet (\mathbf{b} \times \mathbf{c})$, (b) $\mathbf{a} \times (\mathbf{b} \times \mathbf{c})$ and $|\,\mathbf{a} \times (\mathbf{b} \times \mathbf{c})\,|$ and (c) $(\mathbf{a} \times \mathbf{b}) \times \mathbf{c}$ and $|\,(\mathbf{a} \times \mathbf{b}) \times \mathbf{c}\,|$.

Worked Solutions

1(a) Solve for $\mathbf{a} \bullet (\mathbf{b} \times \mathbf{c})$, where

$$\mathbf{a} \bullet (\mathbf{b} \times \mathbf{c}) = \begin{vmatrix} a_x & a_y & a_z \\ b_x & b_y & b_z \\ c_x & c_y & c_z \end{vmatrix} = \begin{vmatrix} 4 & -3 & 6 \\ 2 & 5 & 2 \\ 3 & -1 & 7 \end{vmatrix} = 70 \text{ m}^3.$$

1(b) Solve for $\mathbf{a} \times (\mathbf{b} \times \mathbf{c})$, where

$$\mathbf{a} \times (\mathbf{b} \times \mathbf{c}) = \begin{vmatrix} \mathbf{i} & \mathbf{j} & \mathbf{k} \\ a_x & a_y & a_z \\ b_y c_z - b_z c_y & b_z c_x - b_x c_z & b_x c_y - b_y c_x \end{vmatrix}$$

$$= \begin{vmatrix} \mathbf{i} & \mathbf{j} & \mathbf{k} \\ 4 & -3 & 6 \\ 37 & -8 & 17 \end{vmatrix} = 99\,\mathbf{i} + 290\,\mathbf{j} + 79\,\mathbf{k}\ \text{m}^3.$$

Solve for the magnitude, where

$$|\,\mathbf{a} \times (\mathbf{b} \times \mathbf{c})\,| = \sqrt{99^2 + 290^2 + 79^2} = 316.452\ \text{m}^3.$$

1(c) Solve for $(\mathbf{a} \times \mathbf{b}) \times \mathbf{c}$, where

$$(\mathbf{a} \times \mathbf{b}) \times \mathbf{c} = \begin{vmatrix} \mathbf{i} & \mathbf{j} & \mathbf{k} \\ a_y b_z - a_z b_y & a_z b_x - a_x b_z & a_x b_y - a_y b_x \\ c_x & c_y & c_z \end{vmatrix}$$

$$= \begin{vmatrix} \mathbf{i} & \mathbf{j} & \mathbf{k} \\ -36 & 4 & 26 \\ 3 & -1 & 7 \end{vmatrix} = 54\,\mathbf{i} + 330\,\mathbf{j} + 24\,\mathbf{k}\ \text{m}^3.$$

Solve for the magnitude, where

$$|\,\mathbf{a} \times (\mathbf{b} \times \mathbf{c})\,| = \sqrt{54^2 + 330^2 + 24^2} = 335.249\ \text{m}^3.$$

Vector Algebra Tools Solutions

Using the Triple Products Tool: Enter the given components of vector **a** in the Vector **a** user input fields, the given components of vector **b** in the Vector **b** user input fields and the given components of vector **c** in the Vector **c** user input fields. Then click the **a** • (**b** × **c**) button, the **a** × (**b** × **c**) button and the (**a** × **b**) × **c** button. The answers will be displayed in the output fields as shown.

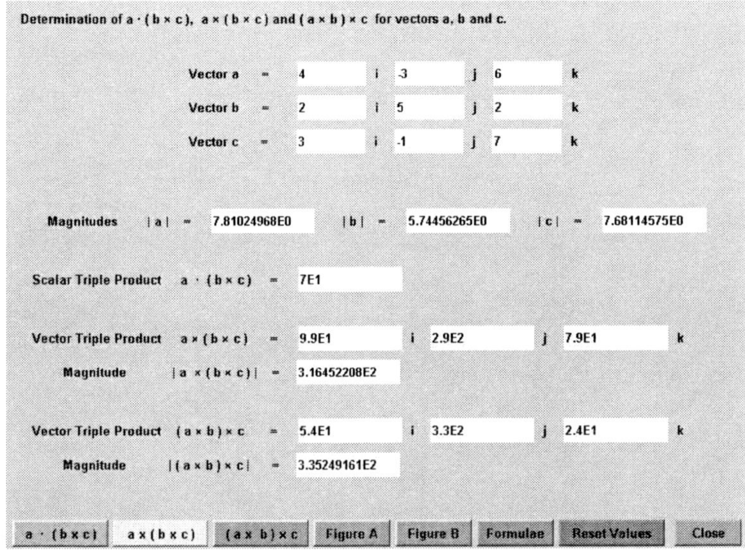

Answers: 1(a) 70 m^3. 1(b) 99 **i** + 290 **j** + 79 **k** m^3 and 316.452 m^3. 1(c) 54 **i** + 330 **j** + 24 **k** m^3 and 335.249 m^3.

Question 2. Determine the volume of a parallelepiped with sides **a** = 4 **i** + 3 **j** + 4 **k**, **b** = 2 **i** − 5 **j** + 4 **k** and **c** = 6 **i** + 4 **j** + 6 **k**. The unit of length is the meter.

Worked Solution

The volume of a parallelepiped whose sides are the vectors **a**, **b**, and **c** is determined by the absolute value of the scalar triple product | **a** • (**b** × **c**) |.

Solve for **a** • (**b** × **c**), where

$$\mathbf{a} \bullet (\mathbf{b} \times \mathbf{c}) = \begin{vmatrix} a_x & a_y & a_z \\ b_x & b_y & b_z \\ c_x & c_y & c_z \end{vmatrix} = \begin{vmatrix} 4 & 3 & 4 \\ 2 & -5 & 4 \\ 6 & 4 & 6 \end{vmatrix} = 4 \text{ m}^3.$$

Hence, the volume of parallelepiped is given by $|\mathbf{a} \bullet (\mathbf{b} \times \mathbf{c})| = 4\ \mathrm{m}^3$.

Vector Algebra Tools Solution

Using the Triple Products Tool: Enter the given components of vector **a** in the Vector **a** user input fields, the given components of vector **b** in the Vector **b** user input fields and the given components of vector **c** in the Vector **c** user input fields. Then click the **a** • (**b** × **c**) button. The answers will be displayed in the output fields as shown.

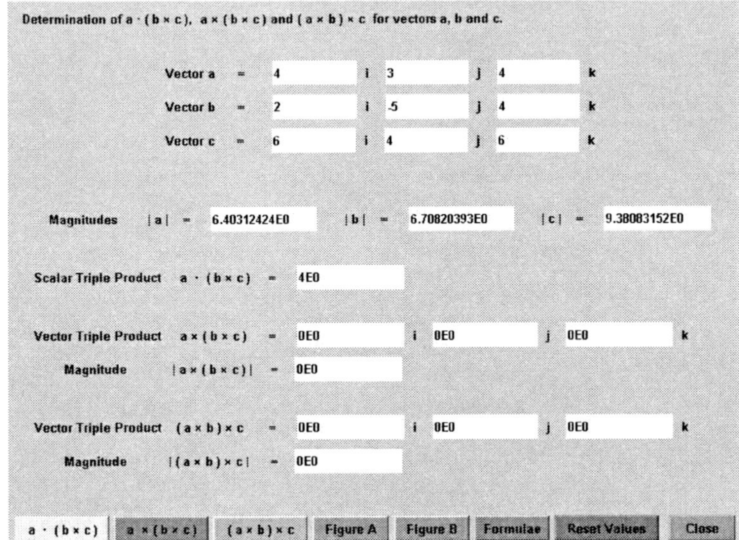

Answer: The volume of the parallelepiped is given by $|\mathbf{a} \bullet (\mathbf{b} \times \mathbf{c})| = 4\ \mathrm{m}^3$.

Question 3. Verify the vector identity expression $\mathbf{a} \times (\mathbf{b} \times \mathbf{c}) = \mathbf{b}(\mathbf{a} \bullet \mathbf{c}) - \mathbf{c}(\mathbf{a} \bullet \mathbf{b})$, where $\mathbf{a} = 2\,\mathbf{i} + 4\,\mathbf{j} + 6\,\mathbf{k}$, $\mathbf{b} = 4\,\mathbf{i} - 8\,\mathbf{j} + 2\,\mathbf{k}$ and $\mathbf{c} = 2\,\mathbf{i} - 6\,\mathbf{j} + 3\,\mathbf{k}$.

Worked Solution

The vector triple product identity can be verified by expanding the given vectors on both sides of the equation into their components.

First solve for $\mathbf{a} \times (\mathbf{b} \times \mathbf{c})$ expressed as the determinant of a 3×3 matrix, where

$$\mathbf{a} \times (\mathbf{b} \times \mathbf{c}) = \begin{vmatrix} \mathbf{i} & \mathbf{j} & \mathbf{k} \\ a_x & a_y & a_z \\ b_y c_z - b_z c_y & b_z c_x - b_x c_z & b_x c_y - b_y c_x \end{vmatrix}$$

$$= \begin{vmatrix} \mathbf{i} & \mathbf{j} & \mathbf{k} \\ 2 & 4 & 6 \\ 12 & -8 & -8 \end{vmatrix}$$

$$= 16\,\mathbf{i} - 56\,\mathbf{j} + 32\,\mathbf{k}.$$

Next solve for the vector identity $\mathbf{b}(\mathbf{a} \bullet \mathbf{c}) - \mathbf{c}(\mathbf{a} \bullet \mathbf{b})$ where,

$\mathbf{a} \bullet \mathbf{c} = (2\,\mathbf{i} + 4\,\mathbf{j} + 6\,\mathbf{k}) \bullet (2\,\mathbf{i} - 6\,\mathbf{j} + 3\,\mathbf{k}) = 4 - 24 + 18 = -2,$

$(\mathbf{a} \bullet \mathbf{c})\,\mathbf{b} = -2(4\,\mathbf{i} - 8\,\mathbf{j} + 2\,\mathbf{k}) = -8\,\mathbf{i} + 16\,\mathbf{j} - 4\,\mathbf{k},$

$\mathbf{a} \bullet \mathbf{b} = (2\,\mathbf{i} + 4\,\mathbf{j} + 6\,\mathbf{k}) \bullet (4\,\mathbf{i} - 8\,\mathbf{j} + 2\,\mathbf{k}) = 8 - 32 + 12 = -12$ and

$(\mathbf{a} \bullet \mathbf{b})\mathbf{c} = -12(2\,\mathbf{i} - 6\,\mathbf{j} + 3\,\mathbf{k}) = -24\,\mathbf{i} + 72\,\mathbf{j} - 36\,\mathbf{k},$

such that,

$$\mathbf{b}(\mathbf{a} \bullet \mathbf{c}) - \mathbf{c}(\mathbf{a} \bullet \mathbf{b}) = (-8\,\mathbf{i} + 16\,\mathbf{j} - 4\,\mathbf{k}) - (-24\,\mathbf{i} + 72\,\mathbf{j} - 36\,\mathbf{k})$$

$$= 16\,\mathbf{i} - 56\,\mathbf{j} + 32\,\mathbf{k}.$$

Therefore,

$$\mathbf{a} \times (\mathbf{b} \times \mathbf{c}) = \mathbf{b}(\mathbf{a} \bullet \mathbf{c}) - \mathbf{c}(\mathbf{a} \bullet \mathbf{b}) = 16\,\mathbf{i} - 56\,\mathbf{j} + 32\,\mathbf{k}.$$

Vector Algebra Tools Solution

Using the Triple Products Tool: Solve for **a** × (**b** × **c**).
Enter the given components of vector **a** in the Vector **a** user input fields, the given components of vector **b** in the Vector **b** user input fields and the given components of vector **c** in the Vector **c** user input fields. Then click the **a** × (**b** × **c**) button. The answer will be displayed in the output fields as shown.

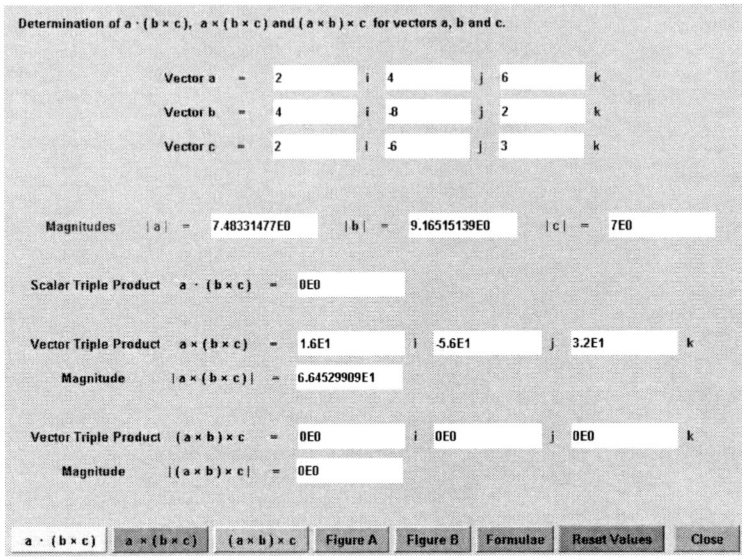

Solution: $\mathbf{a} \times (\mathbf{b} \times \mathbf{c}) = 16\,\mathbf{i} - 56\,\mathbf{j} + 32\,\mathbf{k}$.

Using the Scalar Product Tool: Solve for $(\mathbf{a} \bullet \mathbf{c})$.
Enter the given components of vector **a** in the Vector **a** user input fields and the given components of vector **c** in the Vector **b** user input fields. Then click the Scalar Product **a** • **b** button. The answers will be displayed in the output fields as shown below. Click the Reset Values button before continuing to the next step.

Determination of the scalar product a · b.

Vector a = 2 i 4 j 6 k Scalar multiplier, p = 1
Vector b = 2 i -6 j 3 k Scalar multiplier, q = 1

Magnitude of vector a, | a | = 7.48331477E0 Magnitude of vector b, | b | = 7E0

Vector projection of vector a onto vector b = -8.16326531E-2 i 2.44897959E-1 j -1.2244898E-1 k

Scalar projection of vector a projected onto vector b = -2.857142857E-1
Angle θ between vectors a and b (degrees) = 9.2188E1
Scalar Product a · b = -2E0

Scalar Product a · b | Figure | Formulae | Reset Values | Close

Solution: $(\mathbf{a} \bullet \mathbf{c}) = -2$.

Using the Scalar Product Tool: Solve for (**a** • **b**).
Enter the given components of vector **a** in the Vector **a** user input fields and the given components of vector **b** in the Vector **b** user input fields. Then click the Scalar Product **a** • **b** button. The answers will be displayed in the output fields as shown.

Determination of the scalar product a · b.

Vector a = 2 i 4 j 6 k Scalar multiplier, p = 1

Vector b = 4 i -8 j 2 k Scalar multiplier, q = 1

Magnitude of vector a, | a | = 7.48331477E0 Magnitude of vector b, | b | = 9.16515139E0

Vector projection of vector a onto vector b = -5.71428571E-1 i 1.14285714E0 j -2.85714286E-1 k

Scalar projection of vector a projected onto vector b = -1.309307341E0

Angle Θ between vectors a and b (degrees) = 1.0008E2

Scalar Product a · b = -1.2E1

Scalar Product a · b | Figure | Formulae | Reset Values | Close

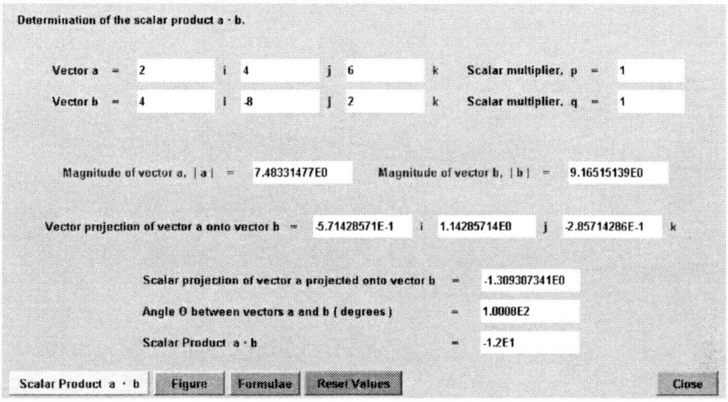

Solution: (**a** • **b**) = −12.

Using the Vector Arithmetic Tool: Solve for (**a** • **c**)**b** − (**a** • **b**)**c**.
Enter the given components of vector **b** in the Vector **a** user input fields and the given components of vector **c** in the Vector **b** user input fields, together with the calculated scalar multipliers of (**a** • **c**) and (**a** • **b**) in the Scalar multiplier p and q user input fields, respectively. Then click the Subtract **a** − **b** button. The answers will be displayed in the output fields as shown.

Determination of a + b, a - b and the scalar multiplcation of vectors a and b.

Vector a =	4	i	-8	j	2	k	Scalar multiplier, p = -2
Vector b =	2	i	-6	j	3	k	Scalar multiplier, q = -12

Addition of a + b =	0E0	i	0E0	j	0E0	k	
| a + b | =	0E0						
Direction cosines:	l =	0	m =	0	n =	0	
Direction angles (degrees):	α =	0	β =	0	γ =	0	

Subtraction of a - b =	1.6E1	i	-5.6E1	j	3.2E1	k	
| a - b | =	6.64529909E1						
Direction cosines:	l =	0.2407717062	m =	-0.8427009716	n =	0.4815434123	
Direction angles (degrees):	α =	76.06790845	β =	147.4264443	γ =	61.21374669	

Add a + b | Subtract a - b | Figure A | Figure B | Formulae | Reset Values | Close

Solution: $\mathbf{b}(\mathbf{a} \bullet \mathbf{c}) - \mathbf{c}(\mathbf{a} \bullet \mathbf{b}) = 16\,\mathbf{i} - 56\,\mathbf{j} + 32\,\mathbf{k}$.

Answer: $\mathbf{a} \times (\mathbf{b} \times \mathbf{c}) = \mathbf{b}(\mathbf{a} \bullet \mathbf{c})\mathbf{b} - \mathbf{c}(\mathbf{a} \bullet \mathbf{b}) = 16\,\mathbf{i} - 56\,\mathbf{j} + 32\,\mathbf{k}$.

6.5 NOTES:

CHAPTER 7

RESULTANT VECTORS IN THE CARTESIAN *X-Y* PLANE

In this chapter the resultant vector of an arbitrary number of concurrent, coplanar vectors in the Cartesian x-y plane is presented. Both the graphical and algebraic determinations of the resultant vector are discussed. Numerous problem solving applications and worked examples are included. Also demonstrated is the use of the Resultant Vector and the Suspended Mass in Equilibrium software tools in solving the example problems.

7.1 DEFINITION AND DETERMINATION OF A RESULTANT VECTOR

7.1.1 Definition of a resultant vector

A **resultant vector** is a single vector that is the **vector sum** of an arbitrary number of vectors having the same units of measurement. For example, in a system comprising of several coplanar forces that act at a common point, the resultant vector is a single force that can replace, with the same effect, the collective forces.

7.1.2 Graphical determination of the resultant vector of two or more coplanar vectors in the x-y plane

Graphically, the resultant vector of any two concurrent, coplanar vectors **a** and **b** in the x-y plane can be determined by employing either the **triangle law** or the **parallelogram law** of vector addition. In each case, the vectors **a** and **b** are represented by their directed line segments, as shown in figure 7.1, where their lengths, drawn to scale, represent the vectors' magnitudes and their arrowheads indicate their directions.

To apply the triangle law of vector addition, a triangle is constructed by first joining the tail of vector **b** to the head of vector **a** and then drawing a vector from the tail of vector **a** to the head of vector **b**, with the vector's arrowhead pointing to the head of vector **b**, thereby completing the third side of the triangle. This third side of the triangle is the resultant vector **r** and is given by $\mathbf{r} = \mathbf{a} + \mathbf{b}$, as shown in figure 7.1(a). Alternatively, applying the parallelogram law of vector addition, a parallelogram is constructed by placing vectors **a** and **b** tail-to-tail and drawing a vector along the diagonal of the parallelogram, representing the resultant vector **r**, as shown in figure 7.1(b). Since vectors **a** and **b** can be translated without changing their magnitudes and directions, both the triangle and parallelogram laws of vector addition are equivalent. Therefore, either method may be used to determine the resultant vector of two coplanar vectors that pass through or act at a common point in the *x-y* plane.

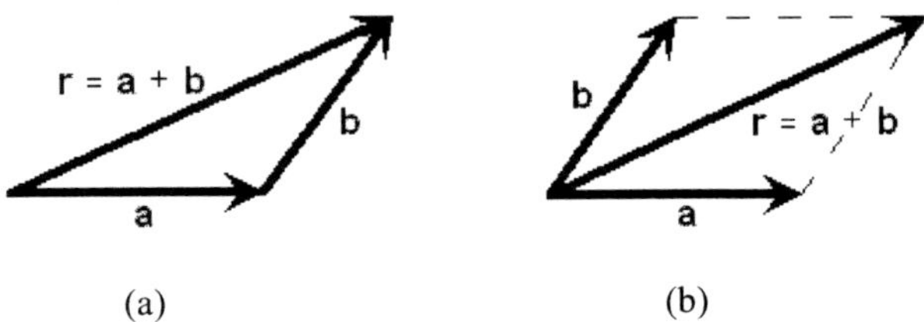

Fig. 7.1. Addition of vectors **a** and **b** using (a) the triangle law and (b) the parallelogram law of vector addition to determine the resultant vector **r**, where $\mathbf{r} = \mathbf{a} + \mathbf{b}$.

To determine the resultant vector of more than two concurrent, coplanar vectors in the *x-y* plane graphically, we use the **polygon method** of vector addition and construct a **polygon of vectors** diagram. The polygon is constructed by drawing each vector to scale and joining the vectors in the appropriate order with the head of the first vector connected to the tail of the second vector, and so on, until all of the vectors have been drawn in their proper directions. The resultant vector is represented by drawing a vector **r** from the tail of the first vector with its arrowhead pointing to the head of the last vector, thereby completing the final side of the polygon, as shown in figure 7.2.

Fig. 7.2. Polygon of vectors diagram of the resultant vector $\mathbf{r} = \mathbf{a}_1 + \mathbf{a}_2 + \mathbf{a}_3 + \cdots + \mathbf{a}_n$.

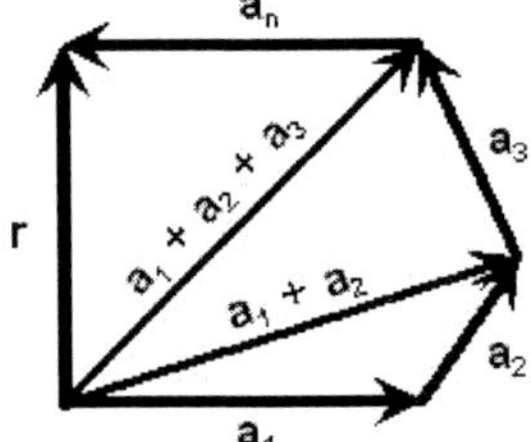

The resultant vector **r** corresponding to the addition of n number of concurrent, coplanar vectors, $\mathbf{a}_1, \mathbf{a}_2, \mathbf{a}_3, \ldots, \mathbf{a}_n$, is given by

$$\mathbf{r} = \mathbf{a}_1 + \mathbf{a}_2 + \mathbf{a}_3 + \cdots + \mathbf{a}_n.$$

It is evident from an inspection of the polygon of vectors diagram, that

$$(\mathbf{a}_1 + \mathbf{a}_2) + \mathbf{a}_3 = \mathbf{a}_1 + (\mathbf{a}_2 + \mathbf{a}_3),$$

and the associative law of vector addition is applicable, as the vectors can be added independent of order without changing their result.

If the addition of three concurrent, coplanar vectors, **a**, **b** and **c**, forms a closed triangle, then the resultant vector **r** is a null vector. Hence,

$$\mathbf{r} = \mathbf{a} + \mathbf{b} + \mathbf{c} = \mathbf{0}.$$

For example, as illustrated in figure 7.3(a), the resultant force $\mathbf{F}_{net}$ of the forces $\mathbf{F}_1$, $\mathbf{F}_2$ and $\mathbf{F}_3$ that act on a particle is zero, and the three force vectors form a closed triangle, as shown in figure 7.3(b). Consequently, any unknown magnitudes and directions of the forces can be determined either graphically if the vectors of the closed triangle are drawn to scale or algebraically from the vector's components based upon the corresponding free-body diagram as shown in figure 7.3(a). Note that, in the case of a closed triangle of forces, any one of the three forces can be considered the **equilibrant force** vector with respect to the other two forces. Since the resultant force is zero, the particle is considered to be in **translational equilibrium** and is either at rest or is moving in a straight line with a constant velocity. In most problem solving applications we only consider particles that are at rest or in static equilibrium.

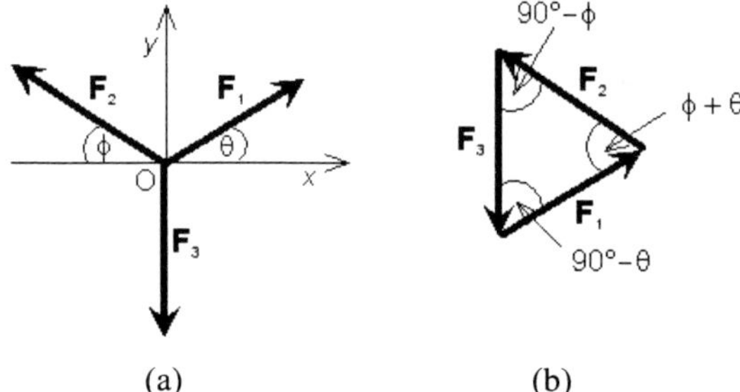

Fig.7.3. The force vectors $\mathbf{F}_1$, $\mathbf{F}_2$ and $\mathbf{F}_3$ acting at the point O (a) are in equilibrium in the free-body diagram and (b) form a closed force triangle when the vectors are drawn in a head-to-tail manner.

Similarly, if the addition of four or more concurrent, coplanar forces acting at a common point forms a closed polygon of vectors, then the resultant force $\mathbf{F}_{net}$ is a null vector, and the system is in **translational equilibrium**.

Hence,

$$\mathbf{F}_{\text{net}} = \mathbf{F}_1 + \mathbf{F}_2 + \mathbf{F}_3 + \mathbf{F}_4 + \cdots + \mathbf{F}_n = \mathbf{0}.$$

7.1.3 Algebraic determination of the resultant vector of two or more coplanar vectors in the *x-y* plane

To determine the resultant vector **r** of the coplanar vectors **a**, **b** and **c** that are expressed in their component form in the *x-y* plane, we separately add the *x* components of vectors **a**, **b** and **c** and then add the *y* components of vectors **a**, **b** and **c**, which are in the directions of the **i** and **j** unit vectors, respectively. For example, if $\mathbf{a} = a_x\,\mathbf{i} + a_y\,\mathbf{j}$, $\mathbf{b} = b_x\,\mathbf{i} + b_y\,\mathbf{j}$ and $\mathbf{c} = c_x\,\mathbf{i} + c_y\,\mathbf{j}$, then the resultant vector **r** can be determined algebraically as follows:

$$\mathbf{r} = \mathbf{a} + \mathbf{b} + \mathbf{c}$$

$$= (a_x\,\mathbf{i} + a_y\,\mathbf{j}) + (b_x\,\mathbf{i} + b_y\,\mathbf{j}) + (c_x\,\mathbf{i} + c_y\,\mathbf{j})$$

$$= (a_x + b_x + c_x)\,\mathbf{i} + (a_y + b_y + c_y)\,\mathbf{j}.$$

Alternatively, a coplanar vector can also be expressed in terms of its magnitude and direction in the *x-y* plane, where the direction is defined by the angle that the vector makes with a specific coordinate axis, which is generally the positive *x*-axis, as discussed in Chapter 3. For vectors expressed in this manner, the resultant vector is determined by first resolving each vector into its *x* and *y* components along the *x* and *y*-axes, respectively. The *x* component of the resultant vector is the algebraic sum of all of the vectors' *x* components, and the *y* component of the resultant vector is the algebraic sum of all of the vectors' *y* components. The resultant vector can then be written in component form as the sum its two vector components, which are defined by its *x* component in the direction of the **i** unit vector and its *y* component in the direction of the **j** unit vector.

This method can be used to determine the resultant vector of any number of coplanar vectors that are expressed in terms of their magnitude and direction in the *x-y* plane. For example, consider of a set of eight arbitrary coplanar vectors, $\mathbf{V}_1$ through $\mathbf{V}_8$, that act on a common point located at the origin in the *x-y* plane, as shown in figure 7.4. Since the vectors act on or pass through the same point in the *x-y* plane, i.e., the point of concurrence, they are also concurrent vectors. The magnitudes of the vectors are given by V_1, V_2, V_3, etc., and φ, ξ, ψ and δ are the angles that vectors $\mathbf{V}_2$, $\mathbf{V}_4$, $\mathbf{V}_6$ and $\mathbf{V}_8$ make with the *nearest* *x*-axis, respectively. Also shown in figure 7.4, is a schematic representation of the corresponding resultant vector **R**, whose direction is determined by the angle θ, i.e., a positive angle in standard position, which by convention is measured in a counterclockwise direction from the positive x-axis.

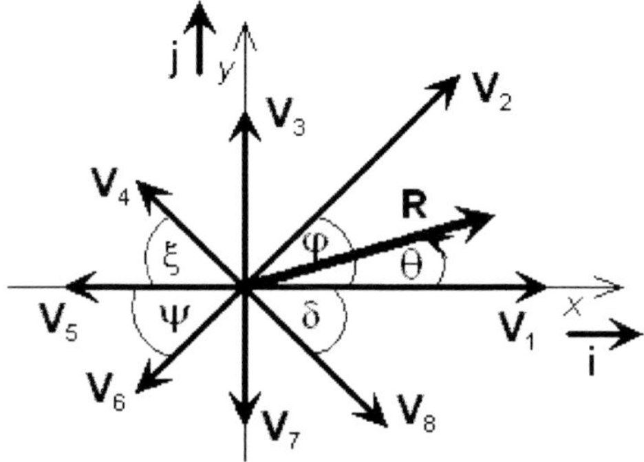

Fig. 7.4. Eight concurrent, coplanar vectors acting on a point located at the origin in the Cartesian x-y plane and their resultant vector **R**.

Referring to figure 7.4, we are able to solve for the resultant vector **R** by first resolving the eight coplanar vectors into their corresponding x and y components that are along the x and y-axes, respectively. Then, by separately adding the vectors' x components and then their y components, we can write

$$R_x = \sum V_x = V_1 + V_2 \cos\varphi - V_4 \cos\xi - V_5 - V_6 \cos\psi + V_8 \cos\delta$$

and

$$R_y = \sum V_y = V_2 \sin\varphi + V_3 + V_4 \sin\xi - V_6 \sin\psi - V_7 - V_8 \sin\delta,$$

where R_x is the algebraic sum of all of the vectors' x components and R_y is the algebraic sum of all of the vectors' y components along the x and y-axes, respectively. Hence, the **resultant vector R** can be expressed in component form as

$$\mathbf{R} = (\sum V_x)\,\mathbf{i} + (\sum V_y)\,\mathbf{j} = R_x\,\mathbf{i} + R_y\,\mathbf{j},$$

where $R_x\,\mathbf{i}$ and $R_y\,\mathbf{j}$ are the vector components or vector resolutes of the resultant vector **R**.

The magnitude of the resultant vector **R** is given by

$$|\mathbf{R}| = R = \sqrt{R_x^{\,2} + R_y^{\,2}}\,.$$

The unit vector $\hat{\mathbf{r}}$ in the direction of the resultant vector **R** is expressed as

$$\hat{\mathbf{r}} = \frac{\mathbf{R}}{|\mathbf{R}|} = \frac{R_x}{|\mathbf{R}|}\mathbf{i} + \frac{R_y}{|\mathbf{R}|}\mathbf{j},$$

where the components of $\hat{\mathbf{r}}$ are the **direction cosines** of **R**, denoted by $[\ell, m]$, and can be expressed as

$$\ell = \cos\alpha = \frac{R_x}{|\mathbf{R}|} \text{ and } m = \cos\beta = \frac{R_y}{|\mathbf{R}|}.$$

The angles α and β are the **direction angles** that the resultant vector **R** makes with the positive x and y-axes, respectively, as shown in figure 7.5.

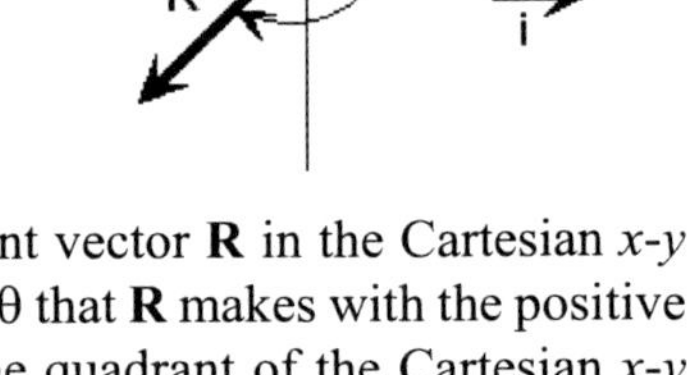

Fig. 7.5. Direction angles α and β of the resultant vector **R**.

Recall that the direction of the resultant vector **R** in the Cartesian x-y plane can also be defined by the angle θ that **R** makes with the positive x-axis. Therefore, depending upon the quadrant of the Cartesian x-y plane in which **R** lies, the angle θ has a value between 0° and 360° and can be determined as follows:

$$\theta = \phi \text{ or } \theta = 180° \pm \phi \text{ or } \theta = 360° - \phi,$$

where the reference angle ϕ is given by $\phi = \tan^{-1}(|\frac{R_y}{R_x}|)$.

If the resultant vector **R** is a null vector, then

$$R_x = \sum V_x = 0 \text{ and } R_y = \sum V_y = 0.$$

For example, where the vectors $\mathbf{V}_1$ through $\mathbf{V}_8$ as shown in figure 7.4 represent the forces that act on a particle such that their resultant force $\mathbf{F}_{\text{net}}$ is zero, the particle is in **translational equilibrium**.

Thus,

$$F_x = \sum V_x = 0 \text{ and } F_y = \sum V_y = 0,$$

where F_x and F_y are the x and y components of the resultant vector $\mathbf{F}_{\text{net}}$, respectively. The condition for translational equilibrium is that there are no net external forces acting on the particle. Therefore, the particle is not accelerated and remains at rest, or if it is moving with a constant velocity in a straight line, it continues to move with the same velocity.

In contrast, the corresponding graphical **polygon of vectors** diagram for determining the resultant vector **R** of the eight concurrent, coplanar vectors is shown in figure 7.6. Specifically, the resultant vector **R** is drawn from the tail of the first vector to the head of the eighth vector and its direction is specified by the angle θ measured from the positive x-axis. However, because of the number of vectors involved and the diagram's complexity, a graphically determination of the resultant vector **R** is not straightforward; therefore, the algebraic approach using the vectors' components is a simpler and more accurate method for determining **R**.

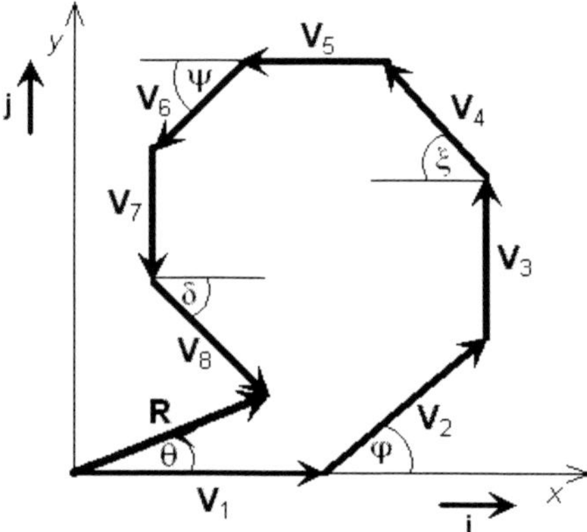

Fig. 7.6. Polygon of vectors diagram of eight concurrent, coplanar vectors in the x-y plane and the resultant vector **R**.

7.2 THE EQUILIBRANT VECTOR

The **equilibrant vector E** is a single vector that, when added to the resultant vector **R**, results in a null vector, such that **R** + **E** = **0**, and **E** = −**R**. The corresponding unit vector in the direction of the equilibrant vector **E** points in the opposite direction of the resultant vector **R** and is given by the unit vector $\hat{\mathbf{e}}$, where $\hat{\mathbf{e}} = -\hat{\mathbf{r}}$. For example, in a system where several coplanar forces act at a common point on a particle, such that the vector sum of the forces is the resultant force $\mathbf{F}_{net}$, then the equilibrant force $\mathbf{F}_E$ is the single force that must be added to $\mathbf{F}_{net}$ to restore the particle to a state of translational equilibrium.

7.3 PROBLEM SOLVING APPLICATIONS OF RESULTANT AND EQUILIBRANT VECTORS

The following examples demonstrate a few of the numerous problem solving applications of resultant and equilibrant vectors in the *x-y* plane.

- Determination of the resultant displacement vector of a particle that has undergone successive displacements from the origin in the *x-y* plane.

- Determination of the resultant or equilibrant force when several coplanar forces act at a common point on a particle in the *x-y* plane.

- Determination of the total linear momentum of a system of particles of an isolated system that collide at a common point in the *x-y* plane. The total linear momentum is the vector sum of the individual momenta of each particle and is given by the expression

$$\mathbf{p}_T = \mathbf{p}_1 + \mathbf{p}_2 + \mathbf{p}_3 + \cdots + \mathbf{p}_n = \text{a constant.}$$

- In electrostatics, the determination of the net electric force or the equilibrant electric force on a point electric charge due to the forces of multiple electric charges located in the *x-y* plane, where the net electric force is given as the vector sum of the individual electric forces on the electric charge. The vector sum of the individual electric forces on the electric charge is known as the **superposition of forces**.

- In electrostatics, the determination of the net electric field or the equilibrant electric field at a point due to the electric fields of multiple electric charges located in the *x-y* plane, where the net electric field is given as the vector sum of the electric fields due to the individual charges. The vector sum of the electric fields due to the individual charges is known as the **superposition of fields**.

7.4 WORKED EXAMPLES

The **Resultant Vector Tool** can be used to determine the resultant and equilibrant vectors in component form when the magnitudes and directions of the vectors in the Cartesian *x-y* plane are known. In addition, the tool can also be used to calculate the magnitudes of the resultant and equilibrant vectors, their direction angles, α and β, and their corresponding unit vectors. See Appendix B for information on how to enter values into any of the Vector Algebra Tools.

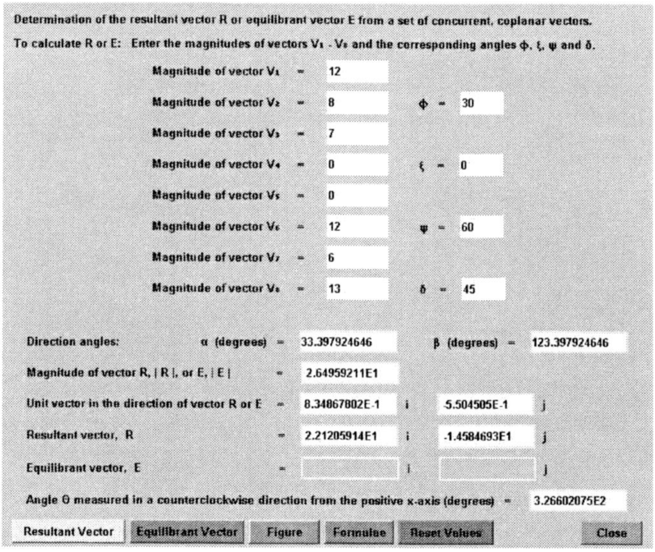

Fig. 7.7. The Resultant Vector Tool from the Vector Algebra Tools Software Program.

The **Suspended Mass in Equilibrium Topic Tool** can be used to determine the tensions in each of the supporting wires holding a suspended mass in equilibrium. See Appendix B for information on how to enter values into any of the Vector Algebra Tools.

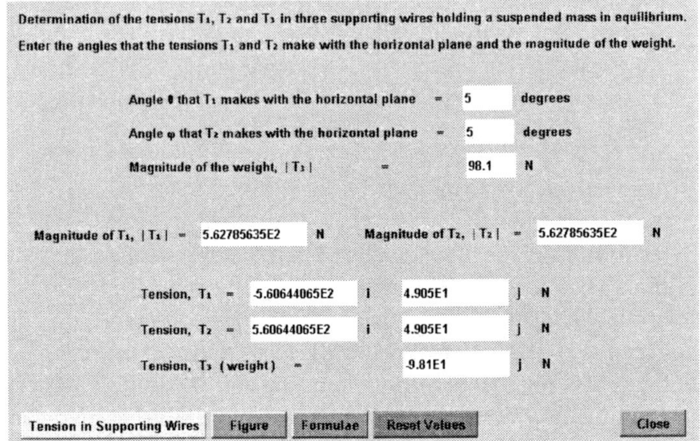

Fig. 7.8. The Suspended Mass in Equilibrium Topic Tool from the Vector Algebra Tools Software Program.

To access the **Suspended Mass in Equilibrium Topic Tool**, open the Resultant Vector menu, and select the **Suspended Mass in Equilibrium Topic Tool**.

Question 1. A cyclist leaves the starting gate of a race and rides 20 km north, 15 km west, 32 km south and 12 km southeast to finish the race, as shown in figure 7.9. Determine the displacement vector **d** and the distance the cyclist is from the starting gate at the end of the race.

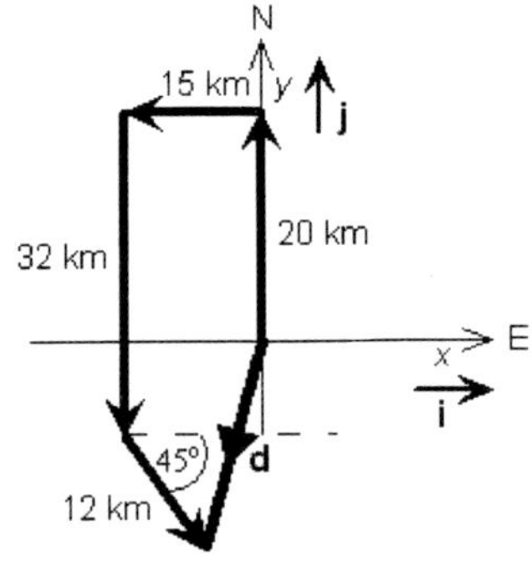

Fig. 7.9.

Worked Solutions

It is convenient to use a coordinate system, as shown in the figure 7.9, where the positive x-axis points to the east in the direction of the **i** unit vector and the positive y-axis points to the north in the direction of the **j** unit vector.

Solve for the resultant displacement vector **d** by first calculating R_x and R_y, the algebraic sums of the x and y components of the individual displacement vectors along the x and y-axes, respectively, where

$$R_x = \sum d_x = -15 + 12 \cos 45° = -6.5147 \text{ km and}$$

$$R_y = \sum d_y = 20 - 32 - 12 \sin 45° = -20.4853 \text{ km.}$$

Hence, the displacement vector **d** of the cyclist from the starting gate at the end of the race, is given by

$$\mathbf{d} = R_x\, \mathbf{i} + R_y\, \mathbf{j} = -6.5147\, \mathbf{i} - 20.4853\, \mathbf{j} \text{ km.}$$

The magnitude of **d**, | **d** |, is the distance that the cyclist is from the starting gate at the end of the race and is given by

$$|\,\mathbf{d}\,| = \sqrt{R_x^{\,2} + R_y^{\,2}} = \sqrt{(-6.5147)^2 + (-20.4853)^2} = 21.496 \text{ km.}$$

Vector Algebra Tools Solutions

Using the Resultant Vector Tool: First click the Figure button and use the illustration to identify the problem's displacement vectors, e.g. $\mathbf{V}_3$, $\mathbf{V}_5$, etc. Close the Figure, and enter each of the given distances traveled by the cyclist in the corresponding user input fields for the Magnitude of vectors $\mathbf{V}_3$, $\mathbf{V}_5$, $\mathbf{V}_7$ and $\mathbf{V}_8$. Next enter the angle of 45° for δ. Enter 0 for all of the remaining angles and magnitudes. Then click the Resultant Vector button. The answers will be displayed in the output fields as shown.

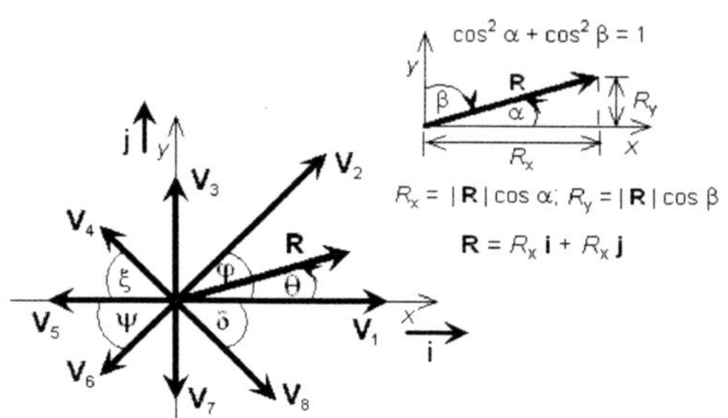

Determination of the resultant vector R or equilibrant vector E from a set of concurrent, coplanar vectors.

To calculate R or E: Enter the magnitudes of vectors V_1 - V_8 and the corresponding angles ϕ, ξ, ψ and δ.

Magnitude of vector V_1 =	0		
Magnitude of vector V_2 =	0	ϕ =	0
Magnitude of vector V_3 =	20		
Magnitude of vector V_4 =	0	ξ =	0
Magnitude of vector V_5 =	15		
Magnitude of vector V_6 =	0	ψ =	0
Magnitude of vector V_7 =	32		
Magnitude of vector V_8 =	12	δ =	45

Direction angles: α (degrees) =	107.641677864	β (degrees) =	162.358322136
Magnitude of vector R, | R |, or E, | E | =	2.14962395E1		
Unit vector in the direction of vector R or E =	-3.03063177E-1 i	-9.52970467E-1 j	
Resultant vector, R =	-6.51471863E0 i	-2.04852814E1 j	
Equilibrant vector, E =	i	j	
Angle θ measured in a counterclockwise direction from the positive x-axis (degrees) =	2.52358322E2		

Resultant Vector | Equilibrant Vector | Figure | Formulae | Reset Values | Close

Answers: The displacement **d** is given by **d** = −6.5147 **i** – 20.4853 **j** km and the distance the cyclist is from the starting gate is | **d** | = 21.496 km.

Question 2. Five concurrent coplanar forces act on an object located at point O, as shown in the figure 7.10. Determine (a) the resultant force vector **F**, (b) its magnitude, | **F** |, and (c) the direction of **F** defined by the angle θ measured in a counterclockwise direction from the positive x-axis.

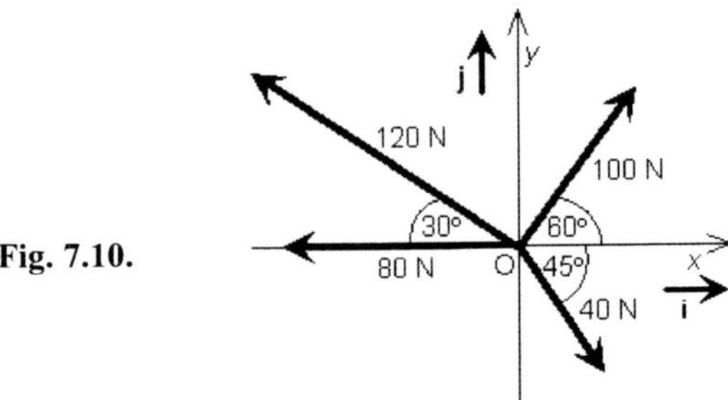

Fig. 7.10.

Worked Solutions

2(a) First calculate the components or resolutes of the resultant force along the x-axis, where R_x is the algebraic sum of all the x components of the forces and is given by

$$R_x = \sum F_x = 100 \cos 60° - 120 \cos 30° - 80 + 40 \cos 45°$$

$$= -105.6388 \text{ N.}$$

Next calculate the components or resolutes of the resultant force along the y-axis, where R_y is the algebraic sum of all the y components of the forces and is given by

$$R_y = \sum F_y = 100 \sin 60° + 120 \sin 30° - 60 - 40 \sin 45°$$

$$= 58.3183 \text{ N.}$$

Hence, the resultant force vector **F** is given by

$$\mathbf{F} = R_x\,\mathbf{i} + R_y\,\mathbf{j} = -105.6388\,\mathbf{i} + 58.3183\,\mathbf{j} \text{ N.}$$

2(b) The magnitude of **F**, | **F** |, is given by

$$|\,\mathbf{F}\,| = \sqrt{R_x^{\,2} + R_y^{\,2}} = \sqrt{(-105.6388)^2 + 58.3183^2} = 120.667 \text{ N.}$$

2(c) The direction of **F** determined by the angle θ is given by

$$\theta = 180° - \phi$$

$$= 180° - \tan^{-1}\left(\left|\frac{R_y}{R_x}\right|\right)$$

$$= 180° - \tan^{-1}\left(\left|\frac{58.3183}{-105.6388}\right|\right)$$

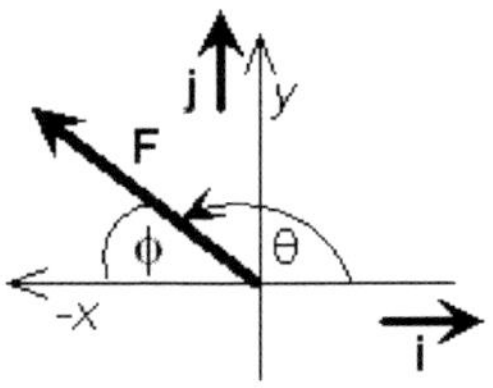

Fig. 7.11.

$= 180° - 28.9° = 151.1°$ (or $\phi = 28.9°$ above the negative x-axis, as shown in figure 7.11).

Note that the angle θ is also given by the direction angle α that the resultant vector **F** makes with positive x-axis and can be determined as follows:

$$\alpha = \cos^{-1}\left(\frac{R_x}{|\mathbf{F}|}\right) = \cos^{-1}\left(\frac{-105.6388}{120.667}\right) = \cos^{-1}(-0.87546) = 151.1°.$$

Vector Algebra Tools Solutions

Using the Resultant Vector Tool: First click the Figure button and use the illustration to identify the problem's force vectors, e.g. $\mathbf{V}_2$, $\mathbf{V}_4$, etc. Close the Figure, and enter the magnitudes for each of the given forces in the corresponding user input fields for the Magnitude of vectors $\mathbf{V}_2$, $\mathbf{V}_4$, $\mathbf{V}_5$, $\mathbf{V}_7$ and $\mathbf{V}_8$. Next enter the given angles for φ, ξ, and δ in the corresponding angle user input fields. Enter 0 for all of the remaining angles and magnitudes. Then click the Resultant Vector button. The answers will be displayed in the output fields as shown.

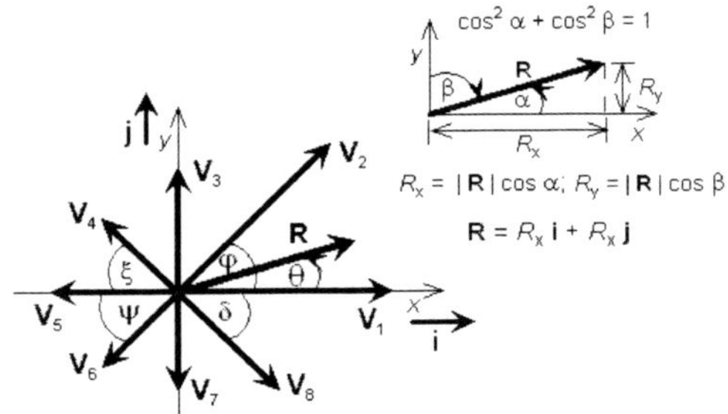

Determination of the resultant vector R or equilibrant vector E from a set of concurrent, coplanar vectors.

To calculate R or E: Enter the magnitudes of vectors V_1 - V_8 and the corresponding angles ϕ, ξ, ψ and δ.

Magnitude of vector V_1 =	0		
Magnitude of vector V_2 =	100	ϕ =	60
Magnitude of vector V_3 =	0		
Magnitude of vector V_4 =	120	ξ =	30
Magnitude of vector V_5 =	80		
Magnitude of vector V_6 =	0	ψ =	0
Magnitude of vector V_7 =	60		
Magnitude of vector V_8 =	40	δ =	45

Direction angles: α (degrees) =	151.098947842		β (degrees) =	61.098947842
Magnitude of vector R, | R |, or E, | E | =	1.20667194E2			
Unit vector in the direction of vector R or E =	-8.75455652E-1	i	4.8329846E-1	j
Resultant vector, R =	-1.05638777E2	i	5.83182691E1	j
Equilibrant vector, E =		i		j
Angle θ measured in a counterclockwise direction from the positive x-axis (degrees) =				1.51098948E2

Resultant Vector | Equilibrant Vector | Figure | Formulae | Reset Values | Close

Answers: 2(a) **F** = −105.6388 **i** + 58.3183 **j** N, 2(b) | **F** | = 120.667 N and 2(c) θ = 151.1° (or ϕ = 28.9° above the negative *x*-axis).

Question 3. An object at point O is held in translational equilibrium by four concurrent forces. If three of the four forces are known, as shown in figure 7.12, determine the magnitude and direction of the fourth force, i.e., the equilibrant force vector **E**.

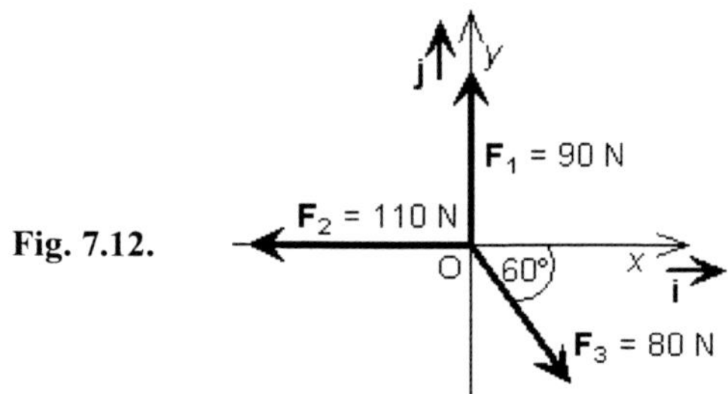

Fig. 7.12.

Worked Solutions

Since the object is in translational equilibrium, the vector sum of *all* the forces acting on the object is given by

$$\sum \mathbf{F} = \mathbf{R} + \mathbf{E} = \mathbf{0},$$

where $\mathbf{R} = \mathbf{F}_1 + \mathbf{F}_2 + \mathbf{F}_3$ and **E** is the equilibrant force vector, $\mathbf{E} = -\mathbf{R}$. First calculate the components or resolutes of the resultant force of the three known forces along the x-axis, where R_x is the algebraic sum of all the x components of the forces and is given by

$$R_x = \sum F_x = -110 + 80 \cos 60° = -70\text{N}.$$

Next calculate the components or resolutes of the resultant force of the three known forces along the y-axis, where R_y is the algebraic sum of all the y components of the forces and is given by

$$R_y = \sum F_y = 90 - 80 \sin 60° = 20.718 \text{ N}.$$

Hence, the resultant vector **R** is given by

$$\mathbf{R} = R_x\, \mathbf{i} + R_y\, \mathbf{j} = -70\, \mathbf{i} + 20.718\, \mathbf{j} \text{ N}.$$

Therefore, the equilibrant vector **E** is given by

$$\mathbf{E} = -\mathbf{R} = 70\, \mathbf{i} - 20.718\, \mathbf{j} \text{ N}.$$

The magnitude of **E**, $|\mathbf{E}|$, is given by

$$|\mathbf{E}| = |-\mathbf{R}| = \sqrt{70^2 + (-20.718)^2} = 73 \text{ N}.$$

The direction of **E** is determined by the angle θ, which is measured in a counterclockwise direction from the positive x-axis, and is given by

$$\theta = 360° - \phi$$

$$= 360° - \tan^{-1}\left(\left|\frac{R_y}{R_x}\right|\right)$$

$$= 360° - \tan^{-1}\left(\left|\frac{-20.718}{70}\right|\right)$$

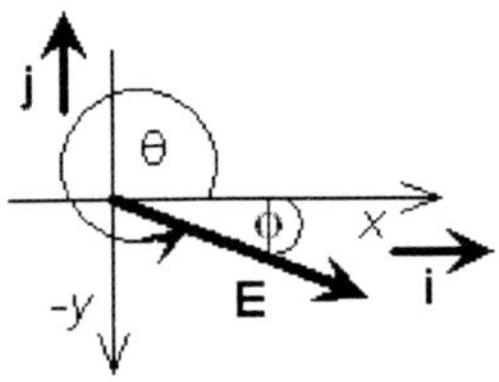

Fig. 7.13.

$$= 360° - 16.49° = 343.51°$$ (or $\phi = 16.49°$ below the positive x-axis, as shown in figure 7.13).

Vector Algebra Tools Solutions

Using the Resultant Vector Tool: First click the Figure button and use the illustration to identify the problem's force vectors, e.g. $\mathbf{V}_3$, $\mathbf{V}_5$, etc. Close the Figure, and enter the magnitudes for each of the given forces in the corresponding user input fields for the Magnitude of vector $\mathbf{V}_3$, $\mathbf{V}_5$ and $\mathbf{V}_8$. Next enter an angle of 60° for δ. Enter 0 for all of the remaining angles and magnitudes. Then click the Equilibrant Vector button. The answers will be displayed in the output fields as shown.

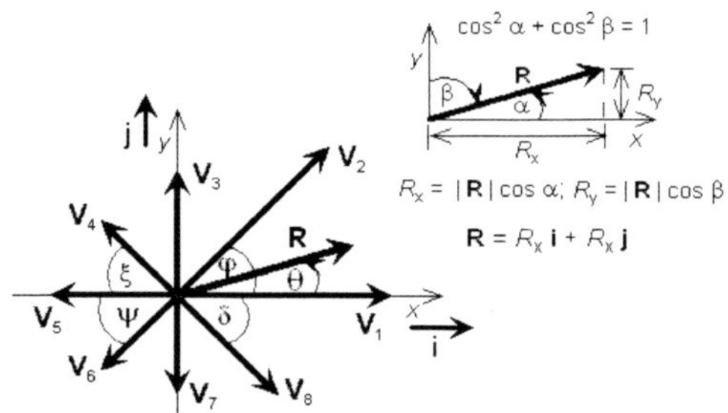

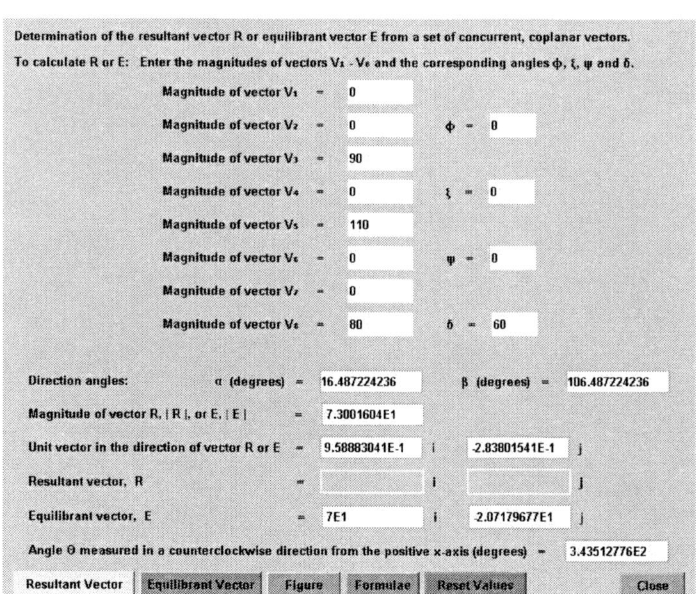

Answers: | **E** | = 73 N and θ = 343.51° (or ϕ = 16.49° below the positive x-axis).

Question 4. Positive point electric charges are place at two vertices of an equilateral triangle, as shown in the figure 7.14. Determine at point P (a) the net electric field vector $\mathbf{E}_{net}$, (b) its magnitude, $|\mathbf{E}_{net}|$, and (c) the direction of $\mathbf{E}_{net}$ defined by the angle θ measured in a counterclockwise direction from the positive x-axis, given that $|\mathbf{E}_1| = 9.55 \times 10^4$ N/C and $|\mathbf{E}_2| = 7.35 \times 10^4$ N/C.

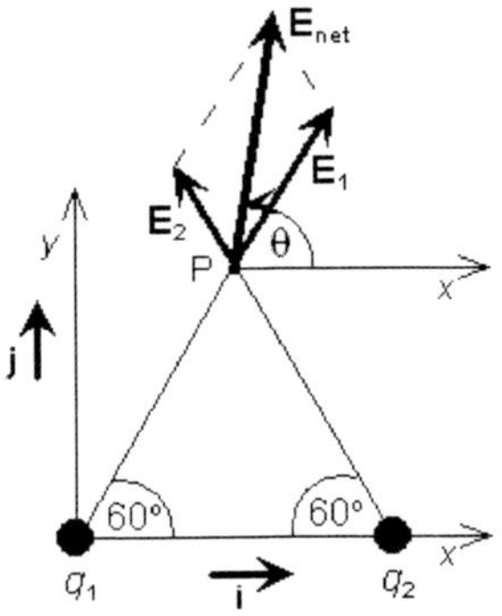

Fig. 7.14.

Worked Solution

4(a) By choosing the coordinate axes as shown in figure 7.14, solve for $\mathbf{E}_{net}$. First calculate the components of the resultant electric field of $\mathbf{E}_1$ and $\mathbf{E}_2$ at point P along the x-axis, where R_x is the algebraic sum of the x components of $\mathbf{E}_{net}$ and is given by

$$R_x = \sum E_x = |\mathbf{E}_1| \cos 60° - |\mathbf{E}_2| \cos 60°$$

$$= 9.55 \times 10^4 \cos 60° - 7.35 \times 10^4 \cos 60°$$

$$= 1.1 \times 10^4 \text{ N/C}.$$

Next calculate the components of the resultant electric field of $\mathbf{E}_1$ and $\mathbf{E}_2$ at point P along the y-axis, where R_y is the algebraic sum of the y components of $\mathbf{E}_{net}$ and is given by

$$R_y = \sum E_y = |\mathbf{E}_1| \sin 60° + |\mathbf{E}_2| \sin 60°$$

$$= 9.55 \times 10^4 \sin 60° + 7.35 \times 10^4 \sin 60°$$

$$= 1.4636 \times 10^5 \text{ N/C}.$$

Hence, the net electric field vector $\mathbf{E}_{net}$ is given by

$$\mathbf{E}_{net} = R_x\,\mathbf{i} + R_y\,\mathbf{j} = 1.1 \times 10^4\,\mathbf{i} + 1.4636 \times 10^5\,\mathbf{j} \text{ N/C}.$$

4(b) The magnitude of $\mathbf{E}_{\text{net}}$, $|\mathbf{E}_{\text{net}}|$, is given by

$$|\mathbf{E}_{\text{net}}| = \sqrt{R_x^{\,2} + R_y^{\,2}}$$

$$= \sqrt{(1.1\times10^4)^2 + (1.4636\times10^5)^2}$$

$$= 1.4677 \times 10^5 \text{ N/C}.$$

4(c) The direction of $\mathbf{E}_{\text{net}}$ is determined by the angle θ and is given by

$$\theta = \tan^{-1}\left(\left|\frac{R_y}{R_x}\right|\right) = \tan^{-1}\left(\left|\frac{1.4636\times10^5}{1.1\times10^4}\right|\right) = 85.7°.$$

Note that the angle θ is also given by the direction angle α that $\mathbf{E}_{\text{net}}$ makes with positive x-axis and can be determined as follows:

$$\alpha = \cos^{-1}\left(\frac{R_x}{|\mathbf{E}_{\text{net}}|}\right) = \cos^{-1}\left(\frac{1.1\times10^4}{1.4677\times10^5}\right) = \cos^{-1}(0.074947) = 85.7°.$$

Vector Algebra Tools Solutions

Using the Resultant Vector Tool: First click the Figure button and use the illustration to identify the problem's vectors, e.g. $\mathbf{V}_2$, $\mathbf{V}_4$. Close the Figure, and enter the magnitudes for each of the given electric fields in the corresponding user input fields for the Magnitude of vector $\mathbf{V}_2$ and $\mathbf{V}_4$. Next enter 60° for the given angles φ, and ξ, in the corresponding angle user input fields. Enter 0 for all of the remaining angles and magnitudes. Then click the Resultant Vector button. The answers will be displayed in the output fields as shown.

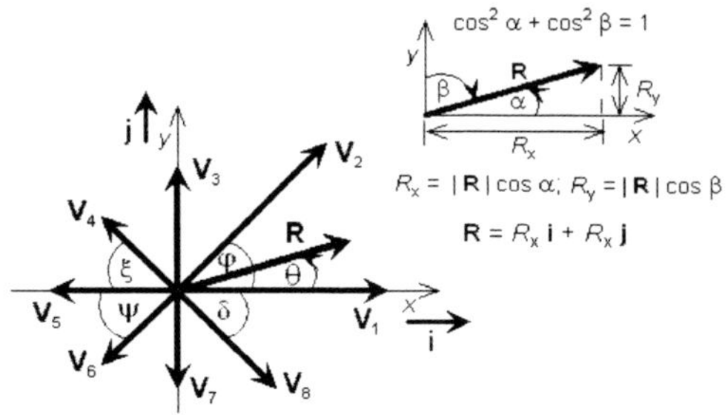

Determination of the resultant vector R or equilibrant vector E from a set of concurrent, coplanar vectors.

To calculate R or E: Enter the magnitudes of vectors V_1 - V_8 and the corresponding angles ϕ, ξ, ψ and δ.

Magnitude of vector V_1 =	0		
Magnitude of vector V_2 =	9.55e4	ϕ =	60
Magnitude of vector V_3 =	0		
Magnitude of vector V_4 =	7.35e4	ξ =	60
Magnitude of vector V_5 =	0		
Magnitude of vector V_6 =	0	ψ =	0
Magnitude of vector V_7 =	0		
Magnitude of vector V_8 =	0	δ =	0

Direction angles:	α (degrees) =	85.701843359		β (degrees) =	4.298156641
Magnitude of vector R, | R |, or E, | E | =		1.4677108E5			
Unit vector in the direction of vector R or E =		7.49466447E-2	i	9.97187545E-1	j
Resultant vector, R =		1.1E4	i	1.46358293E5	j
Equilibrant vector, E =			i		j
Angle θ measured in a counterclockwise direction from the positive x-axis (degrees) =					8.57018434E1

Resultant Vector | Equilibrant Vector | Figure | Formulae | Reset Values | Close

Answers: 4(a) $\mathbf{E}_{net} = 1.1 \times 10^4\, \mathbf{i} + 1.4636 \times 10^5\, \mathbf{j}$ N/C, 4(b) $|\mathbf{E}_{net}| = 1.4677 \times 10^5$ N/C and 4(c) $\theta = 85.7°$.

Question 5. A mass of 25 kg is suspended from a joist by three wires, as shown in the figure 7.15. Determine the tension in each wire. The acceleration due to gravity, $g = 9.81$ m/s^2.

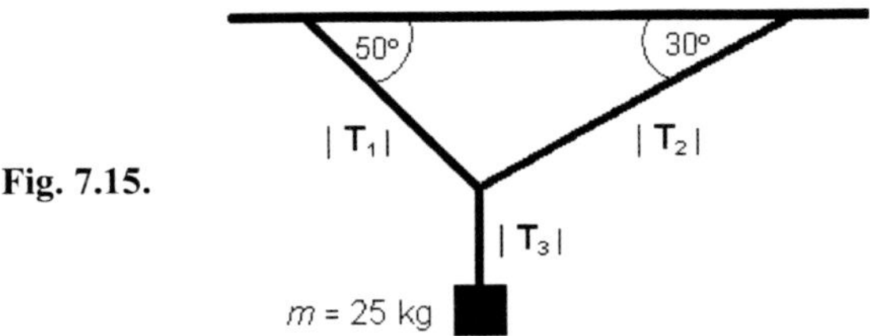

Fig. 7.15.

Worked Solutions

First construct a free-body diagram for the mass, as shown in the figure 7.16, and resolve each of the forces acting on the mass into their x and y components. Note that, since the mass is at rest, i.e., translational equilibrium, the resultant force is zero. Consequently, the x component of the resultant force, $\sum F_x$, is zero and the y component of the resultant

force, $\sum F_y$, is zero. Therefore, using the equations $\sum F_x = 0$ and $\sum F_y = 0$, we can calculate the magnitude of the tensions $|\mathbf{T}_1|$ and $|\mathbf{T}_2|$.
By choosing the coordinate axes, as shown in figure 7.16, first determine the sum of the *x* and *y* components of the forces, i.e., the tensions in the wires, where

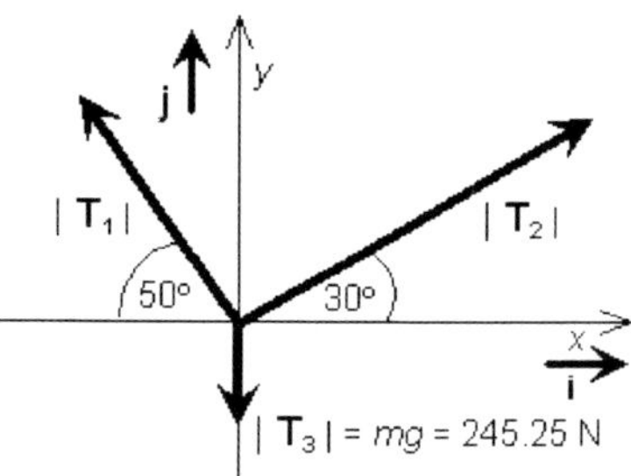

Fig. 7.16.

$$\sum F_x = |\mathbf{T}_1| \cos 50° - |\mathbf{T}_2| \cos 30° = 0 \qquad (1)$$

and

$$\sum F_y = |\mathbf{T}_1| \sin 50° + |\mathbf{T}_2| \sin 30° - |\mathbf{T}_3| = 0. \qquad (2)$$

The magnitude of the tension $|\mathbf{T}_3|$ in the vertical wire is the weight of the mass and is given by

$$|\mathbf{T}_3| = mg = 25 \times 9.81 = 245.25 \text{ N}.$$

By rearranging the terms in equation (1), we can express $|\mathbf{T}_1|$ in terms of $\mathbf{T}_2$ as

$$|\mathbf{T}_1| = \frac{|\mathbf{T}_2| \cos 30^\circ}{\cos 50^\circ}.$$

Then by substituting the above expression for $|\mathbf{T}_1|$ into equation (2), we can solve for $|\mathbf{T}_2|$, where

$$|\mathbf{T}_2| = \frac{|\mathbf{T}_3|}{(\cos 30^\circ \tan 50^\circ) + \sin 30^\circ}$$

$$= \frac{245.25}{(\cos 30^\circ \tan 50^\circ) + \sin 30^\circ}$$

$$= 160.08 \text{ N}.$$

Finally, we can solve for $|\mathbf{T}_1|$ where,

$$|\mathbf{T}_1| = \frac{|\mathbf{T}_2|\cos 30^\circ}{\cos 50^\circ}$$

$$= \frac{160.075\cos 30^\circ}{\cos 50^\circ}$$

$$= 215.67 \text{ N}.$$

Vector Algebra Tools Solutions

Using the Suspended Mass in Equilibrium Topic Tool: Enter the values for the angles θ and φ that the tensions $\mathbf{T}_1$ and $\mathbf{T}_2$ make with the horizontal plane, respectively, and the corresponding magnitude of the weight of the mass. Then click the Tension in Supporting Wires button. The answers will be displayed in the output fields as shown.

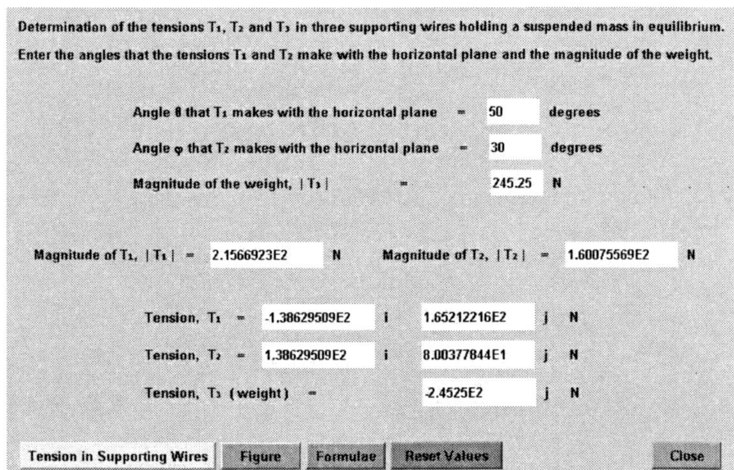

Answers: The magnitudes of the tensions in the three supporting wires are $|\mathbf{T}_1| = 215.67$ N, $|\mathbf{T}_2| = 160.08$ N and $|\mathbf{T}_3| = 245.25$ N.

7.5 NOTES:

APPENDIX A

A Reference Guide to Vector Algebra
Vector Algebra Tools

The Vector Algebra Tools are a comprehensive set of eight easy to use core vector algebra tools that are specialized vector calculators ideally suited for solving vector algebra problems in a Cartesian coordinate system.

The Vector Algebra Tools include the following applications:

1. Vector Arithmetic Tool
2. Vector Composition Tool
3. Scalar Product Tool
4. Vector Product Tool
5. Triple Products Tool
6. 2D-Vector Components Tool
7. 3D-Vector Components Tool
8. Resultant Vector Tool.

Additional features of each tool include:

- Mathematical formulae
- Colorful, annotated figures illustrating the underlying vector concept
- Applicable laws of vector algebra
- Did you know? – a section of important facts and problem solving applications.

The Vector Algebra Tools also includes the following four versatile calculators, each with a companion User's Guide:

- The Scientific Notation SN-205 Calculator™
- The Angular Conversion Calculator
- The Complex Number Arithmetic Calculator
- The Quadratic Formula Calculator.

Also included are the following Vector Algebra Topic Application Tools:

1. Suspended Mass in Equilibrium Topic Tool
2. Forces on an Electric Charge Topic Tool.

Common Vector Algebra Tool Operating Features

Each Vector Algebra Tool has the following operating features:

User input fields:	Boxes in which problem values are entered by the user
Calculation buttons:	Color-coded buttons used to perform the desired calculation
Figure buttons:	Displays an annotated figure illustrating a particular vector concept upon which the tool's calculation is based
Formulae button:	Displays the mathematical equations that are used in the tool's calculations
Reset Values button:	Clears the input and solution field boxes
Close button:	Closes the tool that you are working on and returns you to the main menu screen where you can select another tool to open
Popup menu	By clicking the right mouse button within the frame of the **Vector Algebra Tool**, a popup menu will appear, from which you can open related Vector Algebra Tools and calculators, including the SN-205 Calculator™.

APPENDIX B

Performing a Calculation Using A Reference Guide to Vector Algebra Vector Algebra Tools

Employing the **Vector Arithmetic Tool** as an example, follow these basic steps to perform a calculation using the Vector Algebra Tools.

1. Select the Vector Arithmetic Tool from the main menu of the Vector Algebra Tools.

2. First enter the problem's known values in the corresponding user input fields. Values may be entered in decimal or scientific notation form. To enter a number in scientific notation (i.e., m.d Ex, where m.d is the mantissa and x is the exponent of the power of 10), first enter the value for the mantissa followed by the letter e to designate the exponent and then the value for the power of 10, e.g. 3.675e-3. The International System of Units (abbreviated SI) is used in all calculations. See Appendix E for commonly used SI quantities.

   In the **Vector Arithmetic Tool**, enter the known values for the components of vectors **a** and **b** in the vector **a** and vector **b** user input fields, respectively. Next enter the values for the scalar multipliers p and q corresponding to the vectors **a** and **b** in the scalar multiplier input fields.

3. Determine which calculation operation you need to perform, and click the appropriate calculation button for the solution. For example, in the **Vector Arithmetic Tool**, you can perform vector addition by clicking the Add **a** + **b** button or vector subtraction by clicking the Subtract **a** − **b** button.

   If you have failed to enter a value in a user input field that is required for the calculation or if the value entered is outside the range of calculation, an information message box will appear explaining the nature of the input error.

4. The calculated answer will appear in the solution output field boxes. Many of the tools will also provide additional results. For example, as well as calculating the addition and subtraction of vectors, the **Vector Arithmetic Tool** calculates the magnitude, the direction cosines and the direction angles of the resultant vector. In general, answers will be expressed in decimal form. However, in some tools the answers may be expressed in scientific notation because of the magnitude of the result.

   Should it be necessary to add more than two vectors, simply add the first two vectors and record the answer for the resultant vector. Reset the calculator and then add the next vector to the value for the resultant vector. Repeat this process until all the vectors have been successfully added.

5. An annotated color figure, that graphically illustrates the vector concept upon which the tool's calculation is based, can be viewed and referred to by clicking the **Figure** button.

6. The equations used in the tool's calculations can be viewed and referred to by clicking the **Formulae** button.

7. To solve another equation or if you need to correct your input values, click the **Reset Values** button to clear all of the user input fields, and enter the new set of values.

8. To close a Vector Algebra Tool, simply click the **Close** button, which will return you to the main menu, from which you can select another tool.

APPENDIX C

Vector Identities

Some common vector identities involving scalar and vector triple products for the vectors **a**, **b** and **c**, where **a**, **b** and **c** are nonzero vectors, are summarized in Table C1.

Table C1 Vector identities

1. $\mathbf{a} \bullet (\mathbf{b} \times \mathbf{c}) = \mathbf{b} \bullet (\mathbf{c} \times \mathbf{a}) = \mathbf{c} \bullet (\mathbf{a} \times \mathbf{b}) =$

   $(\mathbf{b} \times \mathbf{c}) \bullet \mathbf{a} = (\mathbf{c} \times \mathbf{a}) \bullet \mathbf{b} = (\mathbf{a} \times \mathbf{b}) \bullet \mathbf{c}$

2. $\mathbf{a} \bullet (\mathbf{b} \times \mathbf{c}) = (\mathbf{b} \times \mathbf{c}) \bullet \mathbf{a} = -\mathbf{a} \bullet (\mathbf{c} \times \mathbf{b})$

   $= -\mathbf{b} \bullet (\mathbf{a} \times \mathbf{c})$

   $= -\mathbf{c} \bullet (\mathbf{b} \times \mathbf{a})$

3. $\mathbf{a} \times (\mathbf{b} \times \mathbf{c}) = (\mathbf{c} \times \mathbf{b}) \times \mathbf{a} = \mathbf{b}(\mathbf{a} \bullet \mathbf{c}) - \mathbf{c}(\mathbf{a} \bullet \mathbf{b})$

   "BAC – CAB" rule

4. $(\mathbf{a} \times \mathbf{b}) \times \mathbf{c} = -(\mathbf{b} \times \mathbf{a}) \times \mathbf{c} = \mathbf{b}(\mathbf{a} \bullet \mathbf{c}) - \mathbf{a}(\mathbf{b} \bullet \mathbf{c})$

   "BAC – ABC" rule

5. $\mathbf{a} \times (\mathbf{b} \times \mathbf{c}) + \mathbf{b} \times (\mathbf{c} \times \mathbf{a}) + \mathbf{c} \times (\mathbf{a} \times \mathbf{b}) = \mathbf{0}$

6. $(\mathbf{a} \times \mathbf{b}) \bullet (\mathbf{c} \times \mathbf{d}) = (\mathbf{a} \bullet \mathbf{c})(\mathbf{b} \bullet \mathbf{d}) - (\mathbf{a} \bullet \mathbf{d})(\mathbf{b} \bullet \mathbf{c})$

7. $(\mathbf{a} \times \mathbf{b}) \times (\mathbf{c} \times \mathbf{d}) = (\mathbf{a} \times \mathbf{b} \bullet \mathbf{d})\mathbf{c} - (\mathbf{a} \times \mathbf{b} \bullet \mathbf{c})\mathbf{d}$

   $= (\mathbf{a} \bullet \mathbf{b} \times \mathbf{d})\mathbf{c} - (\mathbf{a} \bullet \mathbf{b} \times \mathbf{c})\mathbf{d}$

   $= (\mathbf{a} \bullet \mathbf{c} \times \mathbf{d})\mathbf{b} - (\mathbf{b} \bullet \mathbf{c} \times \mathbf{d})\mathbf{a}$

8. $(\mathbf{a} \times \mathbf{b}) \bullet (\mathbf{a} \times \mathbf{b}) = (\mathbf{a} \bullet \mathbf{a})(\mathbf{b} \bullet \mathbf{b}) - (\mathbf{a} \bullet \mathbf{b})^2$

   $= a^2b^2 - (\mathbf{a} \bullet \mathbf{b})^2$

APPENDIX D

Unit Vector Multiplication

The following are the multiplication tables corresponding to the scalar and vector products of the mutually orthogonal, normalized unit vectors **i**, **j** and **k** in a three-dimensional Cartesian coordinate system.

Table D1 Scalar products of the Cartesian unit vectors

$$\mathbf{j} \bullet \mathbf{k} = \mathbf{k} \bullet \mathbf{i} = \mathbf{i} \bullet \mathbf{j} = 0$$

$$\mathbf{i} \bullet \mathbf{i} = \mathbf{i}^2 = 1 \qquad \mathbf{j} \bullet \mathbf{j} = \mathbf{j}^2 = 1 \qquad \mathbf{k} \bullet \mathbf{k} = \mathbf{k}^2 = 1$$

Table D2 Vector products of the Cartesian unit vectors

$$\mathbf{i} \times \mathbf{i} = \mathbf{j} \times \mathbf{j} = \mathbf{k} \times \mathbf{k} = \mathbf{0}$$

$$\mathbf{j} \times \mathbf{k} = \mathbf{i} \qquad \mathbf{k} \times \mathbf{i} = \mathbf{j} \qquad \mathbf{i} \times \mathbf{j} = \mathbf{k}$$

$$\mathbf{k} \times \mathbf{j} = -\mathbf{i} \qquad \mathbf{i} \times \mathbf{k} = -\mathbf{j} \qquad \mathbf{j} \times \mathbf{i} = -\mathbf{k}$$

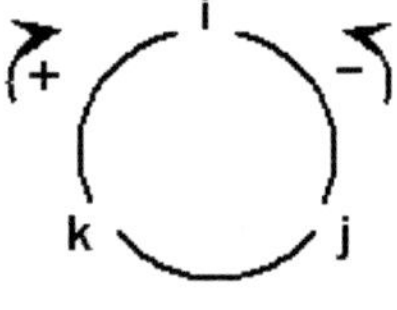

The adjacent illustration can be used to determine the directions of the vector products of any two Cartesian unit vectors. For example, the vector product of **j** × **k** is determined by moving in a clockwise direction from **j** to **k** and then to **i**. On the other hand, the vector product of **k** × **j** is determined by moving in a counterclockwise direction from **k** to **j** to give −**i**, since a counterclockwise circulation results in a negative sign.

APPENDIX E

SI Quantities and Fundamental Physical Constants

The International System of Units (abbreviated as SI) is a system of units that is based on seven fundamental base units that are: the meter (m), the second (s), the ampere (A), the kelvin (k), the kilogram (kg), the mole (mol) and the candela (cd). The physical quantities that are referred to in this handbook correspond to the SI derived units that are expressed algebraically in terms of the base SI units. See Table E1.

Table E1 Commonly used SI derived units

Physical Quantity	Symbol	Unit	Unit Symbol
velocity	$\mathbf{v}$	meter per second	m/s
acceleration	$\mathbf{a}$	meter per second squared	$\mathrm{m/s^2}$
momentum	$\mathbf{p}$	kilogram meter per second	kg m/s
angular momentum	$\mathbf{L}$	kilogram meter squared per second	$\mathrm{kg\ m^2/s}$
force	$\mathbf{F}$	newton	N ($\mathrm{1N = kg\ m/s^2}$)
impulse	$\mathbf{I}$	newton second	N s
torque	$\boldsymbol{\tau}$	newton meter	N m or $\mathrm{kg\ m^2/s^2}$ (1N m = 1J)
energy, work	E, W	joule	J (1J = 1N m)

Table E1 *continued*

Physical Quantity	Symbol	Unit	Unit Symbol
power	P	watt	W (1W = 1J/s)
electric charge	q	coulomb	C (1C = 1A s)
electric field	$\mathbf{E}$	newton per coulomb or volt per meter	N/C or V/m
electric flux	Φ_E	newton meter squared per coulomb	N m^2/C
electric dipole moment	$\mathbf{p}$	coulomb meter	C m
magnetic field	$\mathbf{B}$	tesla	T (1T = 1Wb/m^2)
magnetic flux	Φ_B	weber	Wb (1Wb = 1T m^2)
magnetic dipole moment	$\boldsymbol{\mu}$	ampere meter squared	A m^2
magnetization	$\mathbf{M}$	ampere per meter	A/m
current density	$\mathbf{J}$	ampere per meter squared	A/m^2

Some common fundamental physical constants are listed in Table E2. These constants are expressed in the International System of Units (SI) and are part of the 20 preset physical constant values in the SN-205 calculator™. The values for these physical constants are those recommended by CODATA 2002.

Table E2 Some common fundamental physical constants

Constant	Symbol, Magnitude and Unit
Speed of light in a vacuum	$c = 2.99792458 \times 10^{8}$ m/s
Standard acceleration of gravity	$g = 9.80665$ m/s^2 (adopted value)
Elementary charge	$q = 1.60217653 \times 10^{-19}$ C
Electric constant (permittivity of free space)	$\varepsilon_o = 8.8541878 \times 10^{-12}$ F/m
Magnetic constant (permeability of free space)	$\mu_o = 1.25663706 \times 10^{-6}$ H/m
Electron rest mass	$m_e = 9.1093826 \times 10^{-31}$ kg
Proton rest mass	$m_p = 1.6726217 \times 10^{-27}$ kg
Neutron rest mass	$m_n = 1.6749272 \times 10^{-27}$ kg

Some of the most commonly used SI prefixes that represent powers of 10 are listed in Table E3.

Table E3 Commonly used SI prefixes

Factor	Prefix (symbol)	Factor	Prefix (symbol)
10^9	giga (G)	10^{-3}	milli (m)
10^6	mega (M)	10^{-6}	micro (μ)
10^3	kilo (k)	10^{-9}	nano (n)

Index

A REFERENCE GUIDE TO VECTOR ALGEBRA VECTOR ALGEBRA TOOLS CD – ROM INSTALLATION GUIDE

System Requirements

Supported OS:	Microsoft® Windows® 7, Vista, XP and 2000 operating systems
CPU:	At least a 500 MHz Processor
Memory:	At least 128 MB RAM
Drive:	CD/DVD drive
Hard Disk Space:	At least 70 MB free hard disk space
Display:	Computer screen resolutions: 1024×768, 1280×720, 1280×768, 1280×800, 1280×960, 1280×1024, 1360×768, 1366×768, or 1600×900 pixels.

Installation

To install the Vector Algebra Tools on your computer you must be logged on as the Administrator with your computer display set to one of the computer screen resolutions listed above in the System Requirements. Place the Vector Algebra Tools CD into the CD/DVD tray of the computer, and the Vector Algebra Tools Setup Installation Wizard will start automatically. In Windows Vista or Windows 7, you may be prompted to authorize the installation of the program from an unknown publisher by clicking Allow in Windows Vista or by clicking Yes in Windows 7 in the User Account Control window. The Welcome to the Vector Algebra Tools Setup Wizard will then appear. Follow the on-screen instructions in the Science Tap installation wizard. In the event that the Setup Wizard does not appear automatically:

1. Click the Start button and choose the Run command.

2. Type D:\setup.exe in the text field labeled Open, and click the OK button. If your CD/DVD drive uses a letter other than D, substitute that letter for D.

Running the Vector Algebra Tools

To run the Vector Algebra Tools:

Click the Start button on your desktop and in Windows Vista, XP and 2000 choose Programs and in Windows 7 choose All Programs. Next select Science Tap from the program list and then select the Vector Algebra Tools. If the desktop icon box was checked during the installation wizard, simply double click the Vector Algebra Tools shortcut icon that appears on your computer desktop.

Setting the Computer Screen Resolution

To change the computer display setting on Windows XP or 2000, open Control Panel, double click the Display icon, and, on the Settings tab of the Display Properties window, drag the screen resolution slider to select one of the screen resolutions listed in the System Requirements. Click Apply followed by OK when prompted. To change the computer display settings in Windows Vista, open Control Panel and select the Personalization icon and click on Display Settings. Under resolution on the Monitor tab, drag the resolution slider to select one of the screen resolutions listed in the System Requirements. Then click Apply followed by OK. If you need to change your computer screen resolution in Windows 7, open Control Panel and click Adjust Screen Resolution on the Appearance and Personalization icon. Next, click on the Resolution choice box and drag the resolution slider to select one of the screen resolutions listed in the System Requirements. Then click Apply followed by Keep Changes and then click OK.

Uninstalling the Vector Algebra Tools

To uninstall the Vector Algebra Tools in Windows XP or 2000, open Control Panel and double click the Add or Remove Programs icon. Select the Vector Algebra Tools program from the list of programs. Then click the Change/Remove button to remove and uninstall the Vector Algebra Tools. To uninstall the Vector Algebra Tools in Windows Vista, open Control Panel and double click Uninstall a program within the Programs icon. Select the Vector Algebra Tools program from the list of programs. Then click the Uninstall/Change button to remove and uninstall the Vector Algebra Tools. To uninstall the Vector Algebra Tools in Windows 7, open Control Panel and click the Uninstall a Program within the Programs icon. Select the Vector Algebra Tools program from the list of programs. Then click the Uninstall button to remove and uninstall the Vector Algebra Tools. Alternatively, you may click the Start button on the desktop and in Windows Vista, XP and 2000 choose Programs and in Windows 7 choose All Programs. Then select Science Tap from the program list followed by the Vector Algebra Tools and then click on Uninstall.

END-USER LICENSE AGREEMENT
A REFERENCE GUIDE TO VECTOR ALGEBRA
VECTOR ALGEBRA TOOLS

READ THE FOLLOWING TERMS AND CONDITIONS CAREFULLY BEFORE OPENING THE CD PACKAGE. THIS LEGAL DOCUMENT IS AN AGREEMENT BETWEEN YOU AND JAIN PUBLISHING COMPANY, INC. ("JAIN") AND SCIENCE TAP, INC. ("SCIENCE TAP"), (TOGETHER THE "COMPANIES"). BY OPENING THIS SEALED CD PACKAGE YOU ARE AGREEING TO BE BOUND BY THESE TERMS AND CONDITIONS. IF YOU DO NOT AGREE WITH THESE TERMS AND CONDITIONS, DO NOT OPEN THE CD PACKAGE. RETURN THE **UNOPENED** CD PACKAGE AND ALL ACCOMPANYING ITEMS WITH YOUR PROOF OF PURCHASE TO THE PLACE YOU OBTAINED THEM WITHIN 30 DAYS OF PURCHASE FOR A FULL REFUND.

GRANT OF SINGLE USER LICENSE: In consideration of your purchase of this book and included CD, and your agreement to abide by the terms and conditions of this Agreement, the COMPANIES grant to you a nonexclusive right to use and display the copy of the enclosed SOFTWARE program, "A Reference Guide to Vector Algebra Vector Algebra Tools" by Science Tap, Inc. (hereinafter the "SOFTWARE") on a single computer at a single location so long as you comply with the terms of this agreement. The COMPANIES reserve all rights not expressly granted to you under this Agreement.

OWNERSHIP OF THE SOFTWARE: You own only the magnetic or physical media (the enclosed CD) on which the SOFTWARE is recorded or fixed, but SCIENCE TAP retains all rights, title, and ownership to the SOFTWARE recorded on the original CD copy and all subsequent copies of the SOFTWARE, regardless of the form or media on which the original or other copies may exist. This license is not a sale of the original SOFTWARE or any copy to you.

COPY RESTRICTIONS: This SOFTWARE, including, but not limited to the on-screen text and users guides, and accompanying book (the "Documentation") are subject to copyright. You may **not** copy the Documentation or the SOFTWARE, except that you may make a single copy of the SOFTWARE for backup or archival purposes only. You may be held legally responsible for any copyright infringement that is caused or encouraged by your failure to abide by the terms of this restriction.

USE RESTRICTIONS: The SOFTWARE and Documentation are protected by United States copyright laws and international treaties. You must treat the SOFTWARE and Documentation like any other copyrighted material. You may not copy the Documentation, copy the SOFTWARE except to make one archival or backup copy as provided above, modify or adapt the SOFTWARE or merge it into another program, reverse engineer, disassemble, decompile, adapt, modify or make any attempt to discover the source code of the SOFTWARE, place the SOFTWARE onto a server so that it is accessible via a public network such as the Internet, or otherwise use it on more than one computer or computer terminal at a time, or sublicense, rent, lease, distribute or lend any portion of the SOFTWARE or Documentation to others.

TRANSFER RESTRICTIONS: You may transfer all your rights to use the SOFTWARE and Documentation to another person or legal entity provided you transfer this Agreement, the SOFTWARE and Documentation, including all copies, updates and prior versions to such person or entity and that you retain no copies, including copies stored on computer.

LIMITED WARRANTY AND DISCLAIMER OF WARRANTY: JAIN warrants that for a period of 30 days after delivery of this copy of the SOFTWARE to you, the physical media on which this copy of the SOFTWARE is distributed will be free from defects in materials and workmanship under normal use. Except for the foregoing, the SOFTWARE is provided "AS IS." THE FOREGOING LIMITED WARRANTY IS IN LIEU OF ALL OTHER WARRANTIES OR CONDITIONS, EXPRESS OR IMPLIED, AND THE COMPANIES DISCLAIM ANY AND ALL WARRANTIES OR CONDITIONS, EXPRESSED OR IMPLIED, INCLUDING ANY WARRANTY, DUTY OR CONDITION OF TITLE, NONINFRINGEMENT, MERCHANTABILITY, FITNESS FOR A PARTICULAR PURPOSE,

ACCURACY, CORRECTNESS, COMPLETENESS OR RELIABILITY OF RESPONSES OR RESULTS, REGARDLESS OF WHETHER WE KNOW OR HAD REASON TO KNOW OF YOUR PARTICULAR NEEDS. No employee, agent, dealer or distributor of ours is authorized to modify this limited warranty, or to make any additional warranties. SOME STATES DO NOT ALLOW THE EXCLUSION OF IMPLIED WARRANTIES, SO THE ABOVE EXCLUSION MAY NOT APPLY TO YOU. THIS WARRANTY GIVES YOU SPECIFIC LEGAL RIGHTS, AND YOU MAY ALSO HAVE OTHER RIGHTS THAT VARY FROM STATE TO STATE.

LIMITED REMEDY: The COMPANIES' entire liability and your exclusive remedy shall be the replacement of any CD or other media not meeting our Limited Warranty which is returned to JAIN or to an authorized Dealer or Distributor with a copy of your receipt, or if JAIN or an authorized Dealer or Distributor is unable to deliver a replacement CD or other media that is free of defects in materials or workmanship, you may terminate this Agreement by returning the SOFTWARE and Documentation and your money will be refunded. IN NO EVENT WILL THE COMPANIES BE LIABLE TO YOU FOR ANY DAMAGES, INCLUDING ANY INTERRUPTION OF SERVICE, LOSS OF DATA, LOSS OF CLASSROOM TIME, LOSS OF PROFITS OR ANTICIPATORY PROFITS, LOSS OF SAVINGS, LOSS OF OR IN GRADE(S), GRADE POINTS, GRADE POINT AVERAGES, TEST RESULTS OR TEST SCORES, OR OTHER DIRECT, INCIDENTAL OR CONSEQUENTIAL DAMAGES ARISING FROM THE USE OF OR THE INABILITY TO USE THE SOFTWARE (EVEN IF THE COMPANIES OR AN AUTHORIZED DEALER OR DISTRIBUTOR HAS BEEN ADVISED OF THE POSSIBILITY OF THESE DAMAGES), OR FOR ANY CLAIM BY ANY OTHER PARTY. SOME STATES DO NOT ALLOW THE LIMITATION OR EXCLUSION OF LIABILITY FOR INCIDENTAL OR CONSEQUENTIAL DAMAGES, SO THE ABOVE LIMITATION MAY NOT APPLY TO YOU.

TERM AND TERMINATION: This license is effective until terminated. This license will terminate automatically without notice from the COMPANIES and become null and void if you fail to comply with any provisions or limitations of this license. Upon termination, you shall destroy the Documentation and all copies of the SOFTWARE. All provisions of this Agreement as to warranties, limitation of liability, remedies or damages, and ownership rights shall survive termination.

CONFIDENTIALITY: The SOFTWARE contains trade secrets and proprietary know-how that belongs to SCIENCE TAP and it is being made available to you in strict confidence. ANY USE OR DISCLOSURE OF THE SOFTWARE, OR OF ITS ALGORITHMS, PROTOCOLS OR INTERFACES, OTHER THAN IN STRICT ACCORDANCE WITH THIS LICENSE AGREEMENT, MAY BE ACTIONABLE AS A VIOLATION OF OUR TRADE SECRET RIGHTS.

GENERAL PROVISIONS: This written license agreement is the exclusive agreement between you and the COMPANIES concerning the SOFTWARE and Documentation and supersedes any and all prior oral or written agreements, negotiations or other dealings between you and the COMPANIES. In the event of litigation between you and the COMPANIES concerning the SOFTWARE or Documentation, the prevailing party in the litigation will be entitled to recover attorney fees and expenses from the other party. This license agreement is governed by the laws of the State of California. You agree that the SOFTWARE will not be shipped, transferred or exported into any country or used in any manner prohibited by the United States Export Administration Act or any other export laws, restrictions or regulations.

ACKNOWLEDGMENTS: YOU ACKNOWLEDGE THAT YOU HAVE READ THIS AGREEMENT, UNDERSTAND IT, AND AGREE TO BE BOUND BY ITS TERMS AND CONDITIONS. YOU ALSO AGREE THAT THIS AGREEMENT IS THE COMPLETE AND EXCLUSIVE STATEMENT OF THE AGREEMENT BETWEEN YOU AND THE COMPANIES AND SUPERSEDES ALL PROPOSALS OR PRIOR AGREEMENTS, ORAL OR WRITTEN, AND ANY OTHER COMMUNICATIONS BETWEEN YOU AND THE COMPANIES OR ANY REPRESENTATIVE OF THE COMPANIES RELATING TO THE SUBJECT MATTER OF THIS AGREEMENT. THIS AGREEMENT CANNOT BE MODIFIED OR AMENDED EXCEPT BY A FURTHER WRITTEN INSTRUMENT EXECUTED BETWEEN YOU AND THE COMPANIES.